MINIATURE LIVES

*For Mr Anthony Hiller and Dr Gimme Walter –
my mentors and friends,
who gave me the inspiration to think smart and
the confidence to think big.*

*In loving memory of my dad Colin,
the most skilled fly-catcher I have ever known.*

MINIATURE LIVES

Identifying Insects in Your Home and Garden

Michelle Gleeson

PUBLISHING

© Michelle Gleeson 2016

All rights reserved. Except under the conditions described in the *Australian Copyright Act 1968* and subsequent amendments, no part of this publication may be reproduced, stored in a retrieval system or transmitted in any form or by any means, electronic, mechanical, photocopying, recording, duplicating or otherwise, without the prior permission of the copyright owner. Contact CSIRO Publishing for all permission requests.

National Library of Australia Cataloguing-in-Publication entry

Gleeson, Michelle, author.
Miniature lives : identifying insects in your home and garden/Michelle Gleeson.
9781486301379 (paperback)
9781486301386 (epdf)
9781486301393 (epub)

Includes bibliographical references and index.

Insects – Identification – Handbooks, manuals, etc.
Insect pests – Identification – Handbooks, manuals, etc.
Insects – Life cycles.

595.7

Published by
CSIRO Publishing
Locked Bag 10
Clayton South VIC 3169
Australia

Telephone: +61 3 9545 8400
Email: publishing.sales@csiro.au
Website: www.publish.csiro.au

Front cover: Yellow-banded ichneumon wasp, *Echthromorpha agrestoria* (photograph: James Dorey)
Back cover: Mottled katydid (photograph: Michelle Gleeson)

Set in Minion 9.5/12
Edited by Adrienne de Kretser, Righting Writing
Cover design by Alicia Freile
Typeset by Thomson Digital
Printed in China by 1010 Printing International Ltd

CSIRO Publishing publishes and distributes scientific, technical and health science books, magazines and journals from Australia to a worldwide audience and conducts these activities autonomously from the research activities of the Commonwealth Scientific and Industrial Research Organisation (CSIRO). The views expressed in this publication are those of the author(s) and do not necessarily represent those of, and should not be attributed to, the publisher or CSIRO. The copyright owner shall not be liable for technical or other errors or omissions contained herein. The reader/user accepts all risks and responsibility for losses, damages, costs and other consequences resulting directly or indirectly from using this information.

Original print edition:
The paper this book is printed on is in accordance with the rules of the Forest Stewardship Council®. The FSC® promotes environmentally responsible, socially beneficial and economically viable management of the world's forests.

Foreword

In my early years as a budding naturalist, what wouldn't I have given to have Michelle Gleeson's book at hand to refer to! Life would have been so much easier as I struggled to learn the identities, even just to Family status, of the insects that absorbed so much of my time and attention and, indeed, my growing affection. They were so varied and there were just so many of them – so where to start?

Planet Earth could very well be renamed 'Planet of the Insects'. Today insects comprise more than 80% of all the world's animals – that's more than a million known species with many more yet to be described. As for individual insects, one estimate puts them at a mind-boggling 10 quintillion! I'm not sure what a quintillion is, but it sounds like rather a lot. Anyhow, it is our lot to be involved with insects so it is important that we get to know them. There could be no better introduction than this book.

We humans are the new kids on the block. Insects preceded us by many millions of years and they're basic to our existence as human beings.

The eminent US biologist Ed Wilson has said, 'If the invertebrate animals, including the insects, ever disappeared from the Earth, we humans would be extinct in a few months.' We are all aware of the ongoing threat to the larger animals from habitat destruction. Yet the insects are also under threat and many are already extinct or endangered.

So when we think of conservation it is important not to overlook the vital role the insects have played in the ecology of the planet. Insects have always had a bad press. But directly or indirectly they have had and still have an enormous and mostly positive influence on the way we live. And let's give them credit: only a fraction of 1% of them causes us problems.

The world of insects is a complex one. To ensure its survival we need to become familiar with its highways and byways, and for this we need expert guidance. This book provides that guidance. A dedicated and experienced entomologist, Michelle leads the reader by easy stages down the pathways of insect identification and recognition, behaviour and life stories. While never talking down to her readers she introduces an element of gentle humour and quirkiness into her writing that adds to its attraction and never detracts from its serious purpose. In the right context, entomology can be fun!

The book includes several useful anatomical drawings and excellent close-up photographs that can themselves be used for a quick ID of many common insects. The identification keys you find in most textbooks often cite obscure bits of insect anatomy accessible only from a corpse. The author reinterprets this kind of key, reducing the need for killing bottles and dissection. Instead the reader is given a range of kinder and simpler options to choose from when trying to give a name to an insect. As Michelle points out, we can get important identification clues from an insect's habitat, behaviour, feeding habits and life history. All that is needed is a capacity for patient observation.

My own interest in the invertebrate animals started when I was a child chasing butterflies in an English garden. In the mid 1900s with an English childhood behind me, it was the small inhabitants of an Australian garden that started me on my career as (self-styled) Champion of the Underbug. But I needed a book with some basic knowledge of Australian invertebrates.

In second-hand shops I managed to pick up a few books written for non-scientists by early Australian entomologists. I still treasure those early books and can thank the authors for many happy hours with them, absorbing the natural history of my new country. But my basic knowledge of entomology came only from dry overseas textbooks.

Times have changed and I have now added to my bookshelves a number of attractive, well-produced and readable books written for serious amateurs by Australian entomologists about their own special areas of interest. But up until recently enthusiastic amateurs and beginners still lacked a comprehensive guide to the basic information and linkages they need before continuing their field studies. Now we have Michelle's indispensable guide to insect-watching, for which I am pleased to be writing a foreword.

This book is set to fill an important role, not least in the cause of insect conservation. It will serve as a useful reference work for the serious field entomologist. And hopefully it will help to shatter old prejudices as it introduces new generations to the diverse, secretive, often beautiful and always intriguing minibeasts whose world we so recently came to share.

Densey Clyne
Naturalist, photographer and writer

Contents

	Foreword	*v*
	Acknowledgements	*ix*
	About the author	*x*
1	**Introduction**	**1**
	Why I wrote this book	1
	Approaches to identifying insects	2
	How to use this book	3
2	**Insect basics**	**5**
	What is an insect?	5
	Insect body parts	7
	Taxonomy – how insects are named	11
	Insect Orders – how insects are grouped	12
	Insect growth and lifecycles	20
3	**Morphology – what insects look like**	**25**
	Using this chapter	25
	Tips for getting started	27
	Tools of the trade	27
	The identification key	29
	The little guys – insects less than 5 mm long	64
	Maggots, grubs and caterpillars	64
4	**Habitat – where insects live and occur**	**71**
	Using this chapter	71
	Tips for getting started	72
	Kitchen and pantry	74
	Living room	78
	Bathroom	83
	Bedroom	84
	Around lights	86
	In water	93
	Soil, leaf litter and compost	102
	On trees and shrubs	106
	On citrus trees	116
	In the vegetable garden	120
	On native trees and shrubs	125

In and around the lawn	128
In and around flowers	132
On the bodies of animals – the bloodsuckers	135
In large groups – masses and migrations	138

5 Clever clues – the strange structures and evidence that insects leave behind — 145

Using this chapter	145
Tips for getting started	145
Choose the clue	146
Markings on leaves and bark	146
Lumps and bumps on plants	150
Nests and hideouts	153
Cocoons, cases and eggs	155

6 Insect Orders — 161

How to use this chapter	161
Bees, wasps, ants and sawflies – Order Hymenoptera	162
Beetles – Order Coleoptera	186
Booklice – Order Psocoptera	194
Butterflies and moths – Order Lepidoptera	197
Cockroaches – Order Blattodea	207
Dragonflies and damselflies – Order Odonata	212
Earwigs – Order Dermaptera	219
Fleas – Order Siphonaptera	221
Flies – Order Diptera	225
Grasshoppers and crickets – Order Orthoptera	233
Lacewings – Order Neuroptera	241
Lice – Order Phthiraptera	246
Praying mantids – Order Mantodea	250
Silverfish – Order Thysanura	256
Stick and leaf insects – Order Phasmatodea	259
Termites – Order Isoptera	265
Thrips – Order Thysanoptera	270
True bugs – Order Hemiptera	274
Non-insect arthropods	297

Glossary	*304*
Pronunciation guide	*313*
Bibliography	*316*
Further reading	*321*
Index	*325*

Acknowledgements

Writing an insect identification guide such as this was a huge task and could not have been possible without considerable help. I wish to thank the following people who have been so generous with regards to their time, talent and knowledge. My sincere apologies to anyone I have neglected to thank – this book has been several years in the making and many people have contributed to the final result.

This book has been slowly developing over the years and has been endlessly edited, revised and polished due to the helpful comments from those who kindly took the time to read it – Zoe de Plevitz, Danielle Gleeson, James Gleeson, Beryl Holmes, Gloria Larsen, Ken Lester, Cody Murray and Gimme Walter. Special thanks also go to Mikaela Brusasco, Sarah Clarke, and Kim Pantano and her wonderful volunteers at Osprey House Environmental Centre for testing the insect identification key.

I gratefully acknowledge the photographers of the images used in this book, including Natalie Barnett, Peter Chew, Tony Daley, Vik Dunis, Kathy Ebert, Kristi Ellingsen, John Gooderham and Edward Tsyrlin of *The Waterbug Book*, Anthony and Katie Hiller, Dan Papacek, Erica Siegel and Larena Woodmore. My special thanks also go to James Dorey, who not only provided stunning close-up photographs but took time out of his busy schedule to photograph specimens for this book.

Identifying insects from photographs is a tricky and time-consuming business. My thanks go to Chris Burwell, Peter Cranston, Greg Daniels, Owain Edwards, Ted Edwards, Alexandra Glauerdt, Murray Fletcher, Mark Harvey, Anthony and Katie Hiller, Terry Houston, Christine Lambkin, Melinda McNaught, Mallik Malipatil, Geoff Monteith, Tim New, Dan Papacek, Lindsay Popple, David Rentz, Stefan Schmidt, Adam Slipinksi, Graeme Smith, John Trueman, Edward Tsyrlin, Julianne Waldock, Susan Wright and David Yeates for their expertise.

I am especially grateful to my good friend Julia Toich for taking my scribblings and rantings and turning them into informative illustrations and figures, and to Densey Clyne (I spent many fond hours as a child watching her insect segments on *Burke's Backyard*) for generously contributing the foreword. CSIRO Publishing has been a joy to work with and I am especially thankful to Briana Melideo, Lauren Webb and Tracey Millen for answering my millions of questions (not all of them intelligent).

Last, but not least, I would like to thank my long-suffering family and friends for their love, support and encouragement throughout this journey.

About the author

Michelle Gleeson (*née* Larsen), is known to those around her as 'the bug lady' – no matter how she introduces herself! She is an entomologist and the director and co-founder of Bugs Ed., an educational company that presents a range of hands-on insect workshops throughout Queensland. She is also an Adjunct Industry Fellow at the University of Queensland's School of Biological Sciences, collaborating on biological science outreach programs in remote schools and working on various field research projects.

Michelle's obsession with insects started at an early age. Growing up on the north side of Brisbane, her parents ran a small wildlife refuge from their home, rehabilitating sick and injured wildlife. Michelle's job was to catch grasshoppers and dig up grubs to feed to hungry birds and possums, which resulted in a great deal of time spent outdoors chasing bugs. A brief frisking by her parents would usually turn up an array of critters on her person (some live, some dead, some very squashed). Michelle was around eight when she announced to her family her aspirations to be an entomologist when she grew up. This passion continued throughout her schooling, eventually leading her to university.

Michelle completed a Bachelor of Science majoring in Entomology at the University of Queensland and later received first-class Honours in Entomology. Rather than continuing on the traditional path of academia, Michelle decided to pursue her passion for educating those around her, especially children, about the amazing world of insects. Bugs Ed. was launched in 2003 with her good friends Anthony and Katie Hiller from the Mount Glorious Biological Centre, and by 2005 her business was operating out of its own premises and educating thousands of children about insects across Queensland each year.

Michelle's job as an entomologist is incredibly diverse. One day she might be taking her pet stick insects to visit a class of kindergarten students, the next she is teaching Year 11 biology students how to pin out and display an insect collection. Her summers often involve research projects out in the bush and she regularly films insect-themed stories with children's TV shows. A skilled insect 'wrangler', she has worked on television commercials and film shoots, and has been lucky enough to work on set with one of her idols, Sir David Attenborough.

1
Introduction

Why I wrote this book

If I had a dollar for every insect I have been asked to identify, I would be writing this book perched on the deck of a luxury yacht cruising the islands of the Whitsundays and not, sadly, in the cluttered, cold office under my house. The truth is that there are a lot of insects out there, we encounter them on a daily basis, and most people don't know an aphid from an antlion.

All around Australia, inquisitive kids are chasing and catching insects. Bewildered parents are peering into bug-catchers and sheepishly admitting, 'Sorry Jimmy, I'm not sure what it is.' Many teachers are presented with jam jars and lunchboxes crammed with creepy-crawlies and are not quite sure if they are friend or foe.

Keen gardeners, bush restoration volunteers and just about anybody who has tried starting their own veggie patch has encountered many different types of insects. But just what is defoliating your gardenias, leaving strange lumps and bumps on your wattles, or making a meal out of your basil? On the other hand, which beneficial insects do we want to encourage into our gardens to help pollinate our plants or wage war against pests?

Being able to identify an insect you have found in and around your house (not just your garden) has benefits. It allows us to assess the relative amiability or hostility of a given bug, helping us to decide whether to share our space with it or eject it. With the appropriate background knowledge you, too, can solve the mysteries I face daily, including 'Who has been chewing holes in my cashew nuts?' or 'What has been laying eggs on the eaves of my house?' More excitingly, if you can identify insects you can gain insight into the marvellous diversity of creatures that surrounds us and the miniature lives they lead.

Trying to identify one of our six-legged companions can be daunting (yes, insects do have six legs – but don't worry, we will get to that shortly). You can spend an hour trawling through a field guide trying to match pictures and still not find what you are looking for. The internet is a minefield for incorrectly identified pictures, dubious 'fact sheets' and random pop-up windows for termite control and head lice shampoos.

So here it is, the answer to your prayers – a simple, comprehensive, easy-to-read, identification guide to Australian insects that can be used without needing a microscope, a dictionary of scientific terms or a degree in entomology, which is the technical name for the study of insects. Not only will you be able to identify a given insect but you will also learn what it eats, how it grows, how it protects itself against predators and if it is potentially dangerous or beneficial.

This book is designed as a launch pad for a journey of discovery into the fascinating world of insects. I structured it to be a great starting-point for anyone who is interested in learning more about insect biology and identification. The chapters are chock-full of other useful resources, such as books and keywords for internet searches, to fuel your quest for knowledge.

2 INTRODUCTION

Mottled katydid (*Ephippitytha trigintiduoguttata*).

This book has been specifically designed for anyone to pick up and use, no matter what their level of expertise or where they live in Australia. Gardeners, nature lovers, science or biology students, teachers, parents and grandparents of bug-crazed kids can use this book to identify insects in their home or garden.

Approaches to identifying insects

Different approaches can be taken to identify an insect (this does not generally include lifting your shoe and examining the remains of what just ran past you on the bathroom floor – whole specimens are preferable).

Morphology – what insects look like

The first and most obvious way to identify an insect is based on what it looks like. Is it long and leggy, furry and frumpy, or dainty and delicate? Many insects have gauzy wings resembling delicate lace curtains. Others have chunky legs shaped like a chicken drumstick that are used for jumping. Some even have mouthparts like children's party blowers, for sucking up liquid. By closely examining these weird and wonderful body parts, you may be able to solve the mystery surrounding an insect's identity.

Habitat – where insects live and occur

Some insects are loyal to a particular habitat – the location where an insect naturally lives, feeds or otherwise typically occurs. For example, whirligig beetles live on the water's surface in freshwater creeks and ponds. The caterpillars of certain butterflies only eat the leaves of a single type of plant. Some insects are even attracted by human activities – male rhinoceros beetles in eastern Australia are often found battling each other under the bright lights at service stations on warm summer nights. Becoming familiar with an insect's feeding habits and living arrangements is another approach that can be used to uncover its identity.

Clever clues – the strange structures and evidence that insects leave behind

Some insects operate under the cover of darkness or remain hidden in secret lairs, so cannot be identified based on what they look like. Many are masters of disguise and go undetected, while others are so small that we don't realise they are there. A few insects are so strange in appearance that we fail to recognise them as living creatures. Nevertheless, they may well leave behind clues that are more obvious to us. With the right approach, we can learn to interpret these clues and use them to deduce what insect is in the vicinity, and maybe even

Scientists use the word 'cryptic' to describe the behaviour or appearance of some insects. This means that they are hidden, secretive or difficult to detect.

track it down. Weird egg cases, furry cocoons, and lumps, bumps and bite marks on leaves can all be used to identify an insect that eludes our perception.

How to use this book

Choose an approach to identify your insect
This book allows you to choose, from the approaches above, the most effective way to identify an insect (or its traces) you find in your home or garden. All chapters tie together, so if you are unable to identify an insect using one approach, you can try a different one (Fig. 1.1). For example, you might find an insect crawling on the footpath or wriggling about on the roof of a car. It is unlikely that the insect lives there, so instead of trying to identify it by focusing on its habitat (Chapter 4), head straight to the chapter of this book that deals with morphology (Chapter 3). See Fig. 1.1 for some guidelines.

Learn the basics
Don't know an abdomen from an antenna? Having trouble telling insects apart from other creepy-crawlies? Fear not! 'Insect basics' (Chapter 2) guides you through body parts, classification, lifecycles and other handy information. I strongly recommend you read that chapter first, to arm you with background information you will need to navigate through this book. At the back of the book there is a Glossary which serves as a quick reference to help you understand any unfamiliar words, and a Pronunciation Guide to help you say them like an entomologist.

What you will find in this book
With over 60 000 described species of insects in Australia, it is very difficult to choose which ones to include in an identification book such as this. Instead of giving you a precise guide to a particular insect species, this book allows you to sort insects into major groups. The insects in each group share many features,

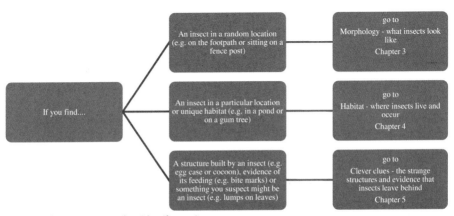

Fig. 1.1: Choose an approach to identify your insect.

including body shape, mode of feeding and means for growth and reproduction, and each such group is called an Order (see pp. 12–13). Information pages on each insect Order are provided in Chapter 6, including details on their general appearance, lifecycle, biology and some amazing facts to wow unsuspecting dinner guests, as well as colour photographs of commonly encountered species as examples.

What you will not find in this book
This book allows you to identify the large, easy to see and commonly encountered Orders of insects (18 Orders). Insect Orders whose members are difficult to see with the naked eye, are very uncommon or live where you are unlikely to encounter them, are not included in this book. See Table 2.7 (pp. 19–20) for a list of them.

Not every creepy-crawly you find in your home or garden is an insect. Chapter 2 helps you to distinguish insects from their close relatives such as spiders and centipedes, creatures I refer to as non-insect arthropods. See the end of Chapter 6 for a bit more information about those.

Encouragement for beginners
Insect diversity is immense, and identifying insects can initially be daunting to the beginner. But don't despair. Once you have worked out what one particular unknown insect is, you have a reference point. Your newfound knowledge makes the next one easier to identify. And once you know that one, you are on the way to building a picture of the insect world. It gets easier as you go. Although there are some real puzzles out there, working hard to crack them (perhaps by using the further reading lists and other sources of information in this book) helps to make it all that much more interesting.

2
Insect basics

This chapter will bring you up to speed with things you need to know about insects, including how to tell them apart from other creepy-crawlies, how we name and group them, how their bodies are designed and the ways in which they grow and change throughout their life.

What is an insect?

Most people unceremoniously label anything with more than four legs a 'bug'. But it is a lot more interesting than that and if you are familiar with a little tricky terminology it will help you work out what you are looking at. The information below and the diagram (Fig. 2.1) should help you to understand exactly what an insect is and how it differs from other organisms.

Insects are invertebrates, that is, an animal that lacks a spine. Invertebrates are a large group that includes many familiar animals such as insects, worms, snails, crabs, leeches and jellyfish. One of the largest and most diverse groups of invertebrates is the arthropods (to which insects belong). Arthropods have a hard outer shell known as an exoskeleton. Slippery and slimy invertebrates mostly lack an exoskeleton, are not arthropods and do not feature in this book. Of course some exceptions do occur, including cherry slugs (the immature stages of a type of wasp, see Chapter 6, p. 186), and these oddballs make insect diversity even more interesting. Arthropods usually have jointed legs to help with locomotion. Insects, spiders, scorpions, millipedes, crayfish and ticks are all examples of arthropods.

Arthropods are further divided into several different groups, based on how many legs they have. Insects form one such group and have the following traits: six legs, three distinct body parts (the head, thorax and abdomen) and a single pair of antennae. Chapter 6 (p. 297) provides a brief summary of some non-insect arthropods we commonly encounter in our homes and gardens.

You may come across the word 'minibeast' in insect books and documentaries and on websites. This is a non-scientific term that is often used to cover any small critters that we see in our homes and garden. While some people use the term minibeast as a substitute for arthropod, others lump other invertebrates such as snails, slugs, leeches and worms under the title.

How many insects are there?

Insects are the most dominant group of organisms on our planet. They account for ~60% of all the species of plants and animals that have been described to date. Well over a million species have been formally named; however, entomologists predict there could be anywhere between 2 million and 50 million extra out there, awaiting discovery. Scientists estimate that at any given time, there are around 10 quintillion individual insects scuttling around our planet. In case you were wondering, that looks like this: 10 000 000 000 000 000 000! In fact, for every person on Earth (around 7 billion) there are around 150 million insects. Some are very common and others are on the brink of extinction.

This display of Christmas beetles (of which there are 35 species in Australia alone) illustrates the strength of insects in both their numbers and diversity.

Bugs versus true bugs

The word 'bug' is used by many as a substitute for 'insect', including myself from time to time! But a true bug belongs to a special Order of insects with piercing and sucking mouthparts, known as Hemiptera (Chapter 6, p. 274). So all 'bugs' are insects, but not all insects are 'bugs'.

Like all true bugs, this juvenile bee killer assassin bug (*Pristhoraneus pluylpennis*) has piercing and sucking mouthparts

Insect body parts

Identifying insects requires some basic knowledge of their morphology, that is, the structure of their body. This section outlines the basic body parts of an insect (see Table 2.1 and Fig. 2.2). Insect groups have different variations of these basic body parts. This allows them to exploit certain habitats and food types, and this is discussed in the information page for each specific Order (Chapter 6).

Types of insect antennae

The shape and size of insect antennae is highly variable. Entomology textbooks have specific names (usually Latin in origin) for the different types of insect antennae. For example, *pectinate* antennae have segments shaped like the teeth in a comb, *moniliform* antennae resemble a string of beads. The figures in Table 2.2 avoid tricky terminology and instead highlight the overall shape of the antennae and their length in relation to the rest of the body. These terms are used throughout the book to describe and compare the antennae of insects in different Orders.

Types of insect mouthparts

Insects have modifications to their mouthparts to enable them to exploit

8 INSECT BASICS

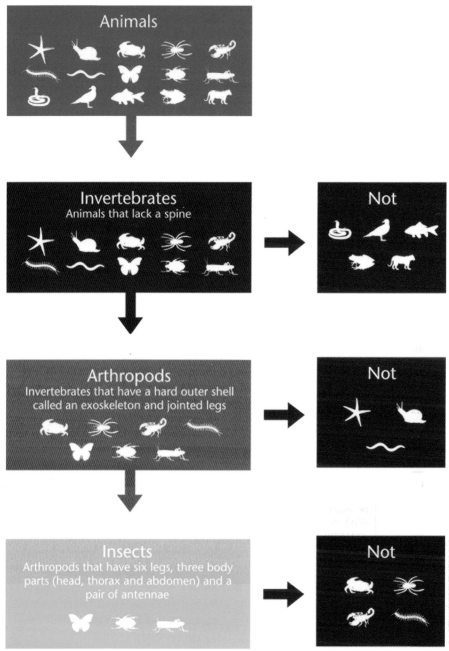

Fig. 2.1: Classification of insects and their relatives.

Table 2.1: Insect body parts

Structure	Description
Head	The front end of an insect's body. This section is in charge of the insect's feeding and sensory systems and bears the eyes, antennae and mouthparts. Insect heads are composed of a series of segments fused together into a solid capsule and can vary greatly in shape and size, depending on how the insect interacts with its environment. See Fig. 2.4 for a diagram of some basic parts of an insect's head.
Antennae (sing. antenna)	Antennae are an insect's alternative to a nose. They are a pair of segmented appendages covered in tiny, highly sensory hairs or *setae* that detect chemicals and other signals from the surrounding environment. See Fig. 2.3 for a close-up view of an insect's antenna and Table 2.2 for some examples of different insect antennae.
Setae (sing. seta)	The tiny hairs on the body of an insect. They are commonly found on the antennae (see Fig. 2.3) and cerci and can help detect smell, taste, touch and vibrations.
Ocelli (sing. ocellus)	Many insects have three ocelli arranged in a triangle on their foreheads, in addition to their two compound eyes (see Fig. 2.4). These are simple eyes which can help an insect to detect levels of light and darkness.
Compound eyes	A single compound eye can be made up of anywhere between a few and tens of thousands of individual facets. These tiny eye-like structures, each with its own lens, join together into a single large eye. Most insects have two compound eyes, although these may be reduced in insects that live underground or in caves, where constant darkness makes vision irrelevant.
Mouthparts	The mouthparts of an insect are very elaborate and are composed of many different parts. Various groups of insects have modifications to these components to help them exploit different food sources. See Table 2.3 for some examples of different insect mouthparts.
Thorax	The middle region of an insect's body. Its main function is locomotion. It is divided into three distinct segments, each bearing a pair of legs. The last two segments may also each provide attachment for a pair of wings (if present at all). The thorax is essentially a rectangular block of very strong muscles to help facilitate the movement of legs and wings.
Legs	Each of the three segments of the thorax bears one pair of legs (the fore, mid and hind legs) and these may be modified to help an insect dig, walk, jump, climb, swim or seize prey. The structure of the leg is much like that of a human, with a coxa (hip joint), trochanter (joint between the hip and thigh), femur (thigh), tibia (shin) and tarsus (foot).
Wings	Many insects have two pairs of wings. The top wings or *fore wings* join to the second of the three segments of the thorax, the bottom or *hind wings* join to the last. Wing structure differs greatly between insect Orders. See Table 2.4 for some examples of different insect wings.
Abdomen	The bottom end of an insect's body. The abdomen is more soft and flexible than the other body segments. It contains the bulk of the circulatory and digestive systems, as well as the reproductive organs. It is usually divided into 11 segments. See Fig. 2.5 for a diagram of some basic parts of an insect's abdomen.

(continued)

Table 2.1: (Continued)

Structure	Description
Spiracles	Tiny oval-shaped openings, known as spiracles, penetrate both sides of the abdominal segments and sometimes the thorax. These connect to a network of air tubes called trachea, which allows the insect to breathe. Insects use muscles to open and close their spiracles during gas exchange.
Cerci (sing. cercus)	A pair of small, finger-like appendages present at the end of some insects' abdomens (see Fig. 2.5). These are covered in tiny sensitive hairs and are used to help an insect detect aspects of its surroundings. In some insect Orders, they may also be used to assist in mating or for defence.

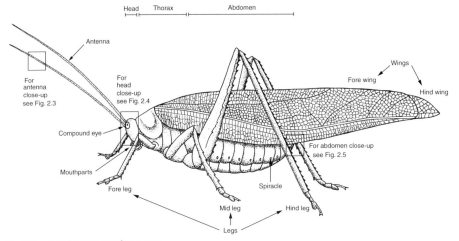

Fig. 2.2: Basic body parts of an insect.

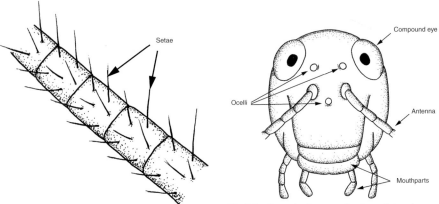

Fig. 2.3: Close-up of an insect's antenna.

Fig. 2.4: Some basic parts of an insect's head.

INSECT BASICS 11

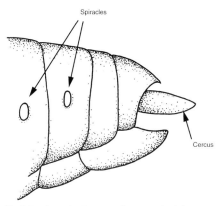

Fig. 2.5: Some basic parts of an insect's abdomen.

different food sources. Strong jaws help to cut and crush food, while needle-like syringes, squishy sponges and curly drinking straws help insects to feed on liquids. Table 2.3 shows simplified diagrams of different insect mouthparts and outlines the modes of feeding of some common insects.

Types of insect wings

Insects demonstrate a variety of wing designs and modifications to enable them to move about in different habitats. Hardened wing covers protect delicate hind wings for a life underground or underwater; some insect groups have ditched their hind wings to improve speed and manoeuvrability. Table 2.4 shows some different types of insect wings.

Taxonomy – how insects are named

Scientists classify insects and give them long names, but there is a good reason for this. The long names are unique and give us a way in which to group insects, first into a small group of very similar insects, then into bigger groups that are somewhat more

Table 2.2: Some types of insect antennae

Type of antennae	Description	Picture
Thread-like (e.g. cockroach)	• Antennae are extremely long and thin and gradually taper from the base to the tip • Antennae often 'slicked back' over the insect's body • Length of antennae is often much longer than the body of the insect • Seen in cockroaches, booklice, crickets	
Long and thin (e.g. wasp)	• Antennae are long, thin and otherwise appear unmodified • Length of antennae is shorter than the body of the insect • Seen in praying mantids,* stick insects, grasshoppers, silverfish	
Bristle-like (e.g. dragonfly)	• Antennae are reduced to hair-like bristles • Difficult to see with the naked eye • Seen in dragonflies, cicadas, some flies	

(continued)

Table 2.2: (Continued)

Type of antennae	Description	Picture
Feathery (e.g. moth)	• Antennae are covered on either side by slender projections, making them look like a feather • Seen in moths	
Clubbed (e.g. lacewing)	• Antennae are gradually thickened at the tip farthest from the head • Seen in butterflies, some lacewings	
Elbowed (e.g. ant)	• Antennae are bent at a 90° angle in the middle • Seen in ants, bees	

* 'Mantids' is commonly used as a plural for 'mantis'.

diverse but still have features in common. This system is *taxonomy*, the scientific practice of classifying, identifying and naming organisms.

All of the world's animals are classified according to the Linnaean system, developed by a famous Swedish naturalist, Carolus Linnaeus, in 1758. His hierarchical system consists of seven main categories, known as taxa, which are ranked in order of increasing specificity. These taxa are sometimes divided into more specific groups, for example, Subphyla, Superfamilies, Subspecies etc. This system allocates two names to every described animal – the generic name (Genus) and the more specific name (Species). For example, the scientific name for the Cairns birdwing butterfly is *Ornithoptera euphorion*. Table 2.5 gives an example of how this butterfly would typically be classified.

Insect Orders – how insects are grouped

Insects are divided into various Orders – that is, groups of insects that share features including body shape, mode of feeding and means for growth and reproduction. All insects fall into one of around 30 Orders. You will be familiar with some of these – for

INSECT BASICS 13

Table 2.3: Some types of insect mouthparts

Type of mouthpart	Description	Picture
Chewing (e.g. beetle)	• Pair of large jaws (mandibles) open and close like scissors and are used to cut, tear, crush and chew food • Seen in beetles, caterpillars, crickets, grasshoppers, dragonflies, earwigs, termites, praying mantids and cockroaches	(front view)
Sponging (e.g. house fly)	• Long tube-like appendage (labium) ending in a pair of fleshy lobes lined with tiny channels to help draw up fluid • Seen in many flies	(side view)
Piercing and sucking (e.g. true bug)	• Pair of sharp, needle-like probes that form two pipes within a piercing and sucking tube, bundled together in a sheath • Collectively called a rostrum • Seen in true bugs (aphids, cicadas and stink bugs) and mosquitoes	(side view)
Sucking (e.g. butterfly)	• A long hollow drinking straw consisting of two tubes locked together by tiny hooks • Seen in butterflies, moths, bees and some flies	(side view)

example, beetles belong to one Order, cockroaches to another. This section outlines the features of the insect Orders most commonly encountered in your home and garden, and introduces you to some obscure Orders not covered in this book.

14 INSECT BASICS

Table 2.4: Some types of insect wings

Type of wing	Description	Picture
Elytra (e.g. beetle)	• Fore wings hardened into shell-like covers (elytra) that sit tight and flat over the membranous hind wings • Fore wings are opaque and hind wings are usually transparent • Seen in beetles and earwigs	
Hemelytra (e.g. crusader bug)	• Part or all of the fore wings are hardened into leathery covers (hemelytra) that sit tight and flat over the membranous hind wings • Fore wings are either partially or wholly opaque and hind wings are usually transparent • Seen in true bugs (members of Hemiptera, Suborder Heteroptera)	
Haltere (e.g. fly)	• Only one pair of wings • Fore wings are usually large and transparent • Hind wings are reduced to tiny club-shaped knobs (halteres) • Seen in flies	
Tegmina (e.g. praying mantis)	• Fore wings form a loose cover over the membranous hind wings • Fore wings are shorter, thicker and more opaque than the hind wings, which are usually transparent • Seen in grasshoppers, crickets, cockroaches, stick insects and praying mantids	
Membranous (e.g. termite)	• Both the fore and hind wings are thin and transparent • Seen in bees, wasps, ants, sawflies, booklice, dragonflies, damselflies, lacewings and termites	

Insect Orders and their features

Table 2.6 helps you recognise the defining features of 18 most commonly encountered insect Orders in Australia. The silhouette pictures highlight the key features of the adults while at rest (i.e. the posture in which you would most commonly see them around your home or garden). A summary of their main diagnostic features is listed and these traits are generally visible to the naked eye. Be sure to check Tables 2.1–2.4 again if you are unsure of any of the terms used. Table 2.6 summarises the features of adult insects only. Juvenile insects may be quite different in appearance. The next section explains in detail the way in which young insects grow, develop and differ from adults. In general, insects that undergo incomplete metamorphosis usually resemble the adults in their basic body shape, but lack wings. For these insects, the figures in Table 2.6 are useful in identifying characteristic features, such as their basic body silhouette and the shape of their legs and antennae. Insects that undergo complete metamorphosis start life as grub-like creatures that share few to no features with the adults. To learn more about these insects, refer to 'Maggots, grubs and caterpillars' (Chapter 3, p. 64).

Cairns birdwing butterfly (*Ornithoptera euphorion*). Source: A. Hiller.

Scientists use the term 'sp.' as an abbreviation of 'species' when they're referring to only one species. They use the abbreviation 'spp.' as a plural for more than one species. For ease of reading, we've used 'sp.' throughout this book to refer to any number of species.

Table 2.5: Classification of a Cairns birdwing butterfly

Taxa	Cairns birdwing butterfly
Kingdom	Animalia (animals)
Phylum	Arthropoda (arthropods)
Class	Insecta (all insects)
Order	Lepidoptera (all butterflies and moths)
Family	Papilionidae (swallowtail butterflies, including birdwing butterflies)
Genus (always written in italics, starting with a capital)	*Ornithoptera* (all birdwing butterflies)
Species (always written in italics, starting in lowercase)	*euphorion* (the Cairns birdwing butterfly)

Table 2.6: Insect Orders and their features

Order	Features	Picture
Bees, wasps, ants and sawflies Order Hymenoptera	• A narrow 'waist' between the thorax and abdomen, with the exception of sawflies (which have a uniformly broad body) • Two pairs of membranous wings that are usually hooked together; wings folded flat over abdomen or held alongside body at rest • Fore wings larger than hind wings • Large compound eyes	
Beetles Order Coleoptera	• Heavily armoured bodies • Often round or oval in shape • Fore wings form a shell-like cover that sits tight and flat over the hind wings and abdomen; hind wings are not visible until they fly • Fore wings meet in a straight line down their backs • No cerci (finger-like appendages) at the end of the abdomen	
Booklice Order Psocoptera	• Tiny • Two pairs of membranous wings, held roof-like over abdomen; often with no wings • Bulging compound eyes • Antennae thread-like	
Butterflies and moths Order Lepidoptera	• Wings and bodies covered in furry scales • Two pairs of large wings that are usually solid in colour and not transparent • Large compound eyes • Antennae clubbed (butterflies), or long and thin or feathery (moths)	
Cockroaches Order Blattodea	• Flattened, oval-shaped bodies, often brown • Long, spiky legs • Leathery fore wings loosely cover membranous hind wings; wings sometimes absent • Large compound eyes • Antennae thread-like • Two cerci (finger-like appendages) at the end of the abdomen	
Dragonflies and damselflies Order Odonata	• Elongate, slender bodies • Legs all equal in size • Two pairs of similarly sized blade-like membranous wings, with networks of tiny black veins • Wings held up vertically above the body, or out flat when at rest • Extremely large, bulbous compound eyes that take up most of the head • Antennae bristle-like and hard to see	

INSECT BASICS 17

Table 2.6: (Continued)

Order	Features	Picture
Earwigs Order Dermaptera	• Elongate, flattened bodies • Legs short, stout and all equal in size • Fore wings form a short shell-like cover that sits tight and flat over the hind wings and abdomen; wings sometimes absent • Fore wings meet in a straight line and extend only a short way down the abdomen • Large pair of pincer-like forceps at the end of the abdomen • Antennae long and thin	
Fleas Order Siphonaptera	• Tiny, usually brown • Body laterally compressed (tall and skinny) • Body covered in spines • Long spiky jumping legs • Never have wings • Pair of claws at the end of each leg	
Flies Order Diptera	• Only one pair of membranous wings present; hind wings reduced down to tiny club-shaped knobs • Legs slender, all equal in size • Large compound eyes	
Grasshoppers and crickets Order Orthoptera	• Elongate bodies • Often well camouflaged • Enlarged hind legs for jumping • Leathery fore wings loosely cover membranous hind wings; wings sometimes absent • Large compound eyes • Antennae long and thin (grasshoppers) or thread-like (crickets)	
Lacewings Order Neuroptera	• Elongate, slender bodies • Two pairs of similarly sized blade-like membranous wings, with networks of tiny veins • Wings held roof-like over the body or out away from the body at rest • Large compound eyes • Antennae long and thin or clubbed	
Lice Order Phthiraptera	• Tiny, often brown • Body dorso-ventrally flattened (short and flat) • Never have wings • Legs short and stout with large claws on the ends	

(continued)

Table 2.6: (Continued)

Order	Features	Picture
Praying mantids Order Mantodea	• Elongate bodies with long slender legs • Often well camouflaged • Fore legs enlarged and lined with sharp spines • Leathery fore wings loosely cover membranous hind wings; wings sometimes absent • Triangular head with large compound eyes • Antennae long and thin	
Silverfish Order Thysanura	• Small, carrot-shaped, flattened bodies • Body usually covered in shiny silvery scales • Never have wings • Eyes reduced or absent • Antennae long and thin • Three long thread-like 'tails' at the end of the abdomen	
Stick and leaf insects Order Phasmatodea	• Extremely elongate bodies with long slender legs • Often well camouflaged • Leathery fore wings loosely cover membranous hind wings; wings sometimes absent • Compound eyes quite small • Antennae long and thin	
Termites Order Isoptera	• Small, delicate and soft-bodied • Usually pale brown, cream or yellow • Two pairs of equally-sized wings that are long, blade-like and membranous, folding flat over the abdomen when at rest and extending well past the end of the body • Wings present in reproductive termites, absent in soldier and worker termites	
Thrips Order Thysanoptera	• Tiny, elongate bodies, often black • Two pairs of long narrow wings with a fringe of tiny hairs along their margins; wings sometimes absent	
True bugs (aphids, cicadas and stink bugs) Order Hemiptera	• Highly variable in shape and size • Mouth always modified into a piercing and sucking tube • Wings membranous and held roof-like over the body; or the fore wings can be leathery and fold tight and flat over the hind wings and abdomen; wings absent in some species • No cerci (finger-like appendages) at the end of the abdomen	

Discrepancies in Order names
At times you will pick up another book on insects or do an internet search on a particular type of insect, only to find it has been allocated to an Order different from that listed in this book. Understanding insect diversity requires ongoing research, and as new insects are discovered (both existing species and extinct species identified from fossils) the way we group insects into Orders continually changes. I have chosen Order names that have been most consistently used over time, as opposed to those based on the most recent scientific findings. For example, historically termites have been grouped under the Order Isoptera and cockroaches in the Order Blattodea. Recent research has suggested that termites and cockroaches share a common ancestor and therefore belong to the same Order, Blattodea. As this is a recent reshuffling, almost all books on the subject of termites still use their original classification, Isoptera. For this reason, this book continues to list termites under that name.

A list of any known synonyms for Order names is included in the summary box at the start of the information pages for each Order. For the most up-to-date information

Table 2.7: Insect Orders not included in this book

Reason for omission	Order	Description
Too obscure These insects have obscure lifestyles and live in places where we are unlikely to look	Stylops Order Strepsiptera	• All are internal parasites of other insects • Female spends her entire life within the host's body • Winged males are tiny and short-lived
	Web-spinners Order Embioptera Synonym: Embiidina	• Spend almost their entire lives hidden in tiny silken tubes in cracks in damp soil and rocks, or under the bark of trees
Too small These insects are so small that we have difficulty seeing them with the naked eye	Proturans Order Protura	• All are less than 2 mm long • Usually found hidden in leaf litter and soil
	Diplurans Order Diplura	• Most are less than 5 mm long. While some larger species exist, they are usually hidden in leaf litter and soil
	Springtails Order Collembola	• Most are less than 3 mm long • Usually found hidden in leaf litter and soil
Rarely found We rarely come across these insects – individuals may appear from time to time, but mostly they are uncommon	Bristletails Order Archaeognatha	• Most come out only at night and many live among rocky crevices on coastal cliffs, or hidden in leaf litter
	Scorpion flies Order Mecoptera	• The winged adults hang on the underside of leaves and are rarely encountered

(continued)

Table 2.7: (Continued)

Reason for omission	Order	Description
Aquatic habitats These aquatic insects are very sensitive to water chemistry or have strict habitat requirements and are not likely to be found in typical urban environments	Stoneflies Order Plecoptera Synonym: Filipalpia, Holognatha, Setipalpia	• The aquatic juveniles prefer clear, fresh, fast-running water • Winged adults and their young are very well camouflaged and difficult to spot
	Caddisflies Order Trichoptera	• Winged adults are usually small, drab and well hidden • The aquatic larvae construct tiny, well camouflaged shelters out of sticks, leaves and pebbles
	Mayflies Order Ephemeroptera	• Nymphs prefer unpolluted waters • Winged adults may swarm near water, but live for no more than 24 hours
	Alderflies Order Megaloptera	• The winged adults are short-lived and the aquatic larvae hide under rocks in freshwater streams

on the classification of insect Orders, refer to the Integrated Taxonomic Information System (ITIS) website (see Bibliography for further details).

Obscure insect Orders
There are around 30 insect Orders formally recognised worldwide. Not all are found in Australia. Ice crawlers, for example, are a group of small insects found only in alpine regions of the Northern Hemisphere. Even within the groups we do find in Australia, not all are insects you would encounter on a daily basis.

To make this book as easy to use as possible, obscure insect Orders have been omitted, for reasons justified in Table 2.7. If your attempts at identifying an insect using this book have been unsuccessful, it may be that your critter belongs to an Order not included in this book. Take the time to read carefully through Table 2.7. If any of the descriptions sound similar to the insect you have found, you may have a starting point for further research.

Insect growth and lifecycles

The journey an insect takes from egg to adult is often long, intricate and interesting. New skins must be grown to accommodate an expanding waistline and growing wings require special strategies.

Moulting – how insects grow larger
Insects (and other arthropods) have a hard outer shell called an exoskeleton. While this tough body armour provides protection against enemies and the elements, its rigid, inflexible structure hinders an insect's growth. For this reason, all insects have to shed or *moult* their exoskeleton (a process known as ecdysis). This allows an insect to grow larger and eventually reach an adult size so that it can reproduce.

When the old skin (we commonly use the term 'skin' to refer to an insect's

INSECT BASICS

A male spiny leaf insect (*Extatosoma tiaratum*) emerges from its old skin as a winged adult. Source: A. Hiller.

exoskeleton, particularly during the moulting process) starts to get a little tight, hormones trigger the development of a new exoskeleton underneath the existing one. In the beginning it is thin and very flexible. Once the new skin is formed, the insect splits open its old one and wriggles out. It takes considerable time and care to ensure that all the legs, antennae and other important appendages are carefully removed from the old exoskeleton. If you have ever extracted yourself from a wetsuit, you can sympathise with a moulting insect. Even some internal organs, such as breathing tubes and parts of the gut lining, must be extracted from the old exoskeleton during the moulting process.

The newly revealed exoskeleton is pale and soft at first and needs several hours to dry and harden properly. During this time the insect is particularly vulnerable. For this reason many insects moult their skin under the cover of darkness, when fewer hungry predators are around. Also, at night, relative humidity is generally high and wind is often absent – both factors that assist in the successful removal of the old exoskeleton. Some insects eat their old skin, either as a source of nourishment or maybe to destroy evidence that may reveal their whereabouts to enemies.

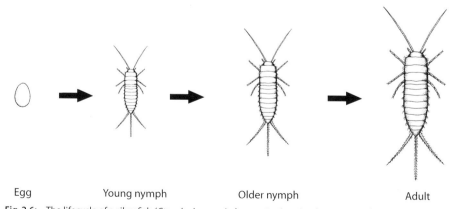

Egg   Young nymph   Older nymph   Adult

Fig. 2.6: The lifecycle of a silverfish (*Ctenolepisma* sp.), demonstrating simple metamorphosis.

The new exoskeleton is slightly larger than the old one, allowing for more room as the insect continues to feed and expand (similar to us sneaking the belt buckle out a notch or two when hitting a buffet restaurant).

Metamorphosis – how insects grow wings
Insects go through a series of changes in their appearance during their journey to adulthood, through a process known as metamorphosis. Most people are familiar with the lifecycle of a caterpillar through to a butterfly, but not all insects grow in this fashion. There are three main ways in which insects grow and develop into adults. These are listed below in terms of their evolutionary development, that is, from the most primitive ways of growing through to the most advanced.

Simple metamorphosis
Simple metamorphosis (also known as ametabolous metamorphosis) is a fairly straightforward way of growing. The juvenile insects or *nymphs* emerge from their eggs as a miniature version of the adult, but lacking reproductive organs. The nymph continues to grow through a series of moults until it reaches sexual maturity, and it continues to moult even as an adult (Fig. 2.6). Simple metamorphosis is possible only in those few ancient insect groups that have never had wings, such as silverfish and bristletails.

Incomplete metamorphosis
The earliest insects with wings also developed in a fairly simple manner, but with an added mechanism for growing wings. Insects that undergo incomplete metamorphosis (also referred to as hemimetabolous metamorphosis) emerge from eggs looking more or less like their parents, but lacking wings and reproductive organs. These young insects (referred to as nymphs) develop wings in external sheaths along their backs. These tiny flaps or *wing buds* increase in size as the insect grows and moults, eventually developing into fully formed wings during the final moult (Fig. 2.7). Once the insect is an adult, it stops moulting.

Incomplete metamorphosis is used by insects such as grasshoppers, cockroaches and stick insects, and you can often see both adults and nymphs living and feeding side by side. This kind of development is also used by insects such as dragonflies that use both aquatic and terrestrial environments.

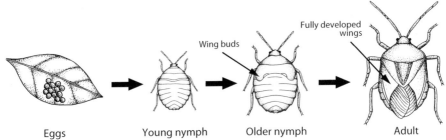

Eggs   Young nymph   Older nymph   Adult

Fig. 2.7: The lifecycle of a bronze orange bug (*Musgraveia sulciventris*), demonstrating incomplete metamorphosis.

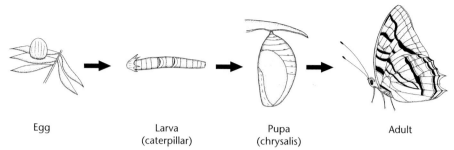

Egg	Larva	Pupa	Adult
	(caterpillar)	(chrysalis)	

Fig. 2.8: The lifecycle of a tailed emperor butterfly (*Charaxes sempronius*), demonstrating complete metamorphosis.

The nymphs (also called larvae or naiads) have adaptations to help them breathe, move and forage underwater, and for this reason look quite different from their land-loving parents. Wing development still happens externally through wing buds, with nymphs exiting the water just before undergoing their final moult into a fully winged adult.

Complete metamorphosis

Complete metamorphosis (also known as holometabolous metamorphosis) is a much more complex type of development and is used by the most recently evolved or *modern* insect groups. Butterflies and moths, beetles, flies, bees, wasps and ants undergo complete metamorphosis.

In this lifecycle, the insect's life is divided into four distinct stages – egg, larva, pupa and adult (Fig. 2.8). The larva (pl. *larvae*) emerges from the egg looking nothing like its winged parents. Also known as caterpillars, maggots or grubs, these juveniles have large, more flexible bodies built for eating, and this is the stage in the lifecycle where most of the growing takes place.

The larva undergoes several moults until it reaches a certain size, which is when special hormones trigger the production of a pupa (pl. *pupae*) under its larval skin. This is revealed upon the final larval moult. Inside the pupa, the larval body is broken down, reorganised and reformed as an adult.

A pupa may be known by several different names, including chrysalis and cocoon, but not all of these are correct. A chrysalis is the special name given to the pupae of butterflies. A cocoon is a bag of silk or other materials, made by the caterpillar, which surrounds and protects the pupa. Cocoons are commonly made by moth larvae, but not by butterflies.

Within the pupa the adult insect puffs up its body, causing a split down its side from which the adult emerges. Adults usually have well-developed wings and more streamlined bodies, and are responsible for dispersal and reproduction.

3
Morphology – what insects look like

Sometimes insects turn up in the strangest places (e.g. hanging from your screen door, clutched in the hands of a toddler, dead on the footpath, crawling around on your garden furniture) and this can make it hard to identify them by their habitat. Fortunately, many groups of insects have consistently characteristic features that help us identify them – this is the focus of this chapter.

Using this chapter

In this chapter you will be identifying your insect using a dichotomous key (here referred to as the identification key), a special tool used by scientists to classify just about anything, for example birds, trees, flowers or seashells. The identification key asks a series of questions about the physical appearance of your insect, gradually discarding possibilities until you are left with a single answer. This answer tells you the group or *Order* that your insect belongs to and refers you to an information page (Chapter 6). There you can learn more about the general feeding habits, lifecycle and survival strategies of the Order your insect belongs to.

The identification key in this book is unique. A lot of traditional dichotomous keys for insects are very complicated and rely on the user being able to turn the insect over to closely examine its mouthparts, or to open up its wings to examine their colour, shape or texture. The identification key developed here uses easily visible characters, such as the shape of the insect's body and limbs. This key enables you to classify an insect with just a photograph, so capturing the critter is not necessary.

How to use the identification key

Start at Question 1 and gradually work your way through the identification key. It is a bit like a choose-your-own-adventure book, with the answer for each question leading you to a different part of the key.

Keep a pen and paper handy and use it to keep track of which path you have taken through the identification key. That way if you make a mistake, you can simply backtrack rather than having to start again from the beginning.

As you look through the identification key, you may notice that some questions appear more than once. This repetition of information is one of the strengths of identification keys such as this, and helps to reinforce the choices you make at each question. Similarly, the answer identifying your insect to a given Order might appear in several parts of the identification key, not just at one unique point. Multiple paths in the key that lead to the same Order simply indicate that its members are highly diverse in their features.

Throughout the identification key you will find symbols to help guide you (see Fig. 3.1). You will also find drawings to help illustrate certain features relating to the questions. The drawing may not exactly match the species of insect you are identifying, so make sure you focus on the features that directly relate to the question.

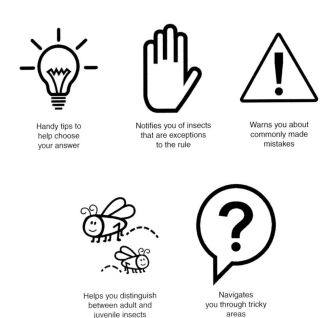

Fig. 3.1: Symbols used in the identification key.

The identification key introduces some basic terminology when referring to the different parts of an insect's body. You can brush up on this terminology in Tables 2.1–2.4 or look it up in the Glossary.

Exceptions and special considerations
The identification key may not be suitable to use when identifying the following types of insects.

Aquatic insects
Insects in aquatic environments such as streams, lakes, ponds and swamps have special adaptations to help them live and breathe underwater, and their basic body shape and structure differs enormously from insects that live on land. For this reason, they have been left out of the identification key, to make it as simple as possible. Should you need to identify an insect you have found in water, refer to Chapter 4, p. 93.

Dead and pinned insects
In the identification key, many of the questions refer to the posture of your insect alive and at rest (e.g. how it folds its wings). For this reason, it can be used for live insects, insect photographs and, to a certain extent, dead insects you may find around your home or garden. You will encounter problems if you try to use the key on dead insects that have been pinned and mounted (a method used by entomologists for displaying specimens in an insect collection). In this case, the wings of an insect are repositioned from their resting posture and opened out to display certain features. If you are a science student or amateur collector using this book to identify insects for a collection, remember that it is common practice to identify your insect to an Order **before** you pin it out.

Tips for getting started

Before you get started, here are some tips that make using this chapter even easier.

Tip 1. Take a photograph

Insects rarely wait patiently on a leaf while you flick through an identification key, and often fly off before you have finished with them. A photograph can be a very useful tool in identifying insects and in this age of digital and mobile phone cameras, it has never been easier to take a photo of your mystery critter. Photographs enable you to see close-up features, recall important characteristics and give you something to compare with the images in this book. Most importantly, a photo saves you having to coax your insect into a jam jar or bug-catcher to study it more closely!

Tip 2. How many legs?

The number of legs is very important in separating insects (which have six legs) from other arthropods (which have more than six), and it is the first question in the identification key. Occasionally you may find a critter that is missing one or more of its limbs. This is often a survival mechanism – a neat trick for escaping predators, much like a lizard losing its tail. This is obviously bad news if you are trying to count the number of legs. So, remember that all arthropods are symmetrical. That is, whatever is present on one side of the body (legs, antennae, wings etc.), is mirrored on the other side.

Tip 3. Exceptions to the rule

Insects form an extremely large and diverse group of organisms. Sometimes, unique types within an insect Order have split off from the rest of the pack and do something a little out of the ordinary. For example, grasshoppers and crickets are characterised by their enlarged hind legs, filled with strong muscles to help them jump. However, an unusual type of cricket, known as stick katydids, do not have enlarged hind legs despite their ability to jump. So, be aware of the occasional oddballs, which are marked with a 'stop hand' symbol (Fig. 3.1) and explained in the identification key.

Tip 4. Got it wrong?

Sometimes you will turn to the Order information page (Chapter 6) for the insect you have just identified, only to discover the pictures and information do not match it. At the start of the information page for each Order, you will find a summary box which outlines the key features of the insect Order and lists other, similar-looking insect Orders you may have it confused with. If you refer to the information pages for these alternative insect Orders, you will find pictures and information that should help you to rectify misidentifications.

Remember, if you are not confident with your attempts at identifying an insect using this identification key, you can use the other identification approaches in this book to confirm your answer. For example, try flicking through the habitat section in Chapter 4 to see if your insect turns up there.

Tools of the trade

Here is a list of optional equipment that makes using the identification key as easy and as accurate as possible.

- Magnifying glass – to make it easier to see the features used in the identification key. A simple hand-held magnifying glass (available from hardware stores, variety stores, chemists or bookstores) is suitable. Hand lenses (also called jeweller's loupes) can be folded up and

tucked into your pocket and are handy for looking at small, slow-moving insects, insect structures (e.g. egg cases) and evidence of feeding (e.g. skeletonised leaves) while outdoors. A head magnifier (a pair of spectacle-type magnifying glasses attached to an adjustable headband) provides an effective, hands-free way of viewing insect specimens or photographs – but may attract some strange looks from friends and family. All magnifiers mentioned can be purchased from online suppliers of entomology equipment.
- Notepad and pencil – to keep a record of the path you take through the identification key.
- Ruler – for accurately measuring the size of your insect (needed for some of the questions in the identification key).
- Camera or mobile phone – to take a photo of your mystery insect.
- A bug-catcher, aquarium or jar – if you can't get a clear photograph of your mystery insect, you can gently coax it into a container for closer examination. Remember to make sure there is plenty of ventilation and keep the container in the shade so that the insect does not suffocate, overheat or drown in droplets of condensation.
- Fine paintbrush – handy for gently coaxing specimens into containers for observation, without hurting them.
- Measuring magnifier – available from online suppliers of entomology equipment, this handy gadget combines a magnifier, ruler and specimen container in one. It is a small plastic container that allows you to capture a specimen for observation and measure its size with the aid of gridlines in the bottom. It also features a magnifying lens in the lid.

The identification key

Question 1. Is it an insect?
Not everything you find scuttling around your home or garden is an insect (see Chapter 2). Insects belong to a big group of animals called arthropods, which includes their non-insect cousins the spiders, scorpions, ticks, mites, centipedes, millipedes, crabs and other crustaceans. Many of these superficially resemble insects.

If your critter has six legs
Go to Question 2

If your critter has more than six legs
It is not an insect, go to Chapter 6, p. 297

It can be tricky to count legs. Some butterfly and moth larvae have leg-like suction cups called prolegs on their bodies that may be confused with true legs. If you count these 'false' legs, you may confuse insect larvae with non-insect groups. Figure 3.8 (p. 66) outlines the difference between insect larvae and similar-looking millipedes and centipedes.

Insects are the only arthropods that have wings. If your critter has wings, skip counting the legs and go straight to Question 2.

What if it has fewer than six legs? All arthropods have **at least** six legs. The only four-legged creatures in your home or garden would be things like lizards and frogs, which are not arthropods.

What if it has no legs?! The young or *larval* stages of some insects are legless. However, they will grow six legs when they are adults, so if your critter has no legs, go to Question 2.
Be careful – legless organisms can be very difficult to identify. Slugs and worms (not covered in this book) can be confused with legless insect larvae, such as maggots. If you can't find your critter in the section 'Maggots, grubs and caterpillars' (p. 64), it is probably not an insect.

Be careful – some spiders mimic insects, some insects mimic spiders. Never assume something is an insect simply because you think it looks like one. Always count the legs carefully.

Question 2. Little or large?
Some really tiny insects belong in their own unique Orders.

Is your insect smaller than this?

If your insect is small – less than 5 mm long or no bigger than the insect shown
Go to 'The little guys' (p. 64)

If your insect is more than 5 mm long or bigger than the insect shown
Go to Question 3

 Not sure how small your insect really is? Compare it to the pictures in 'The little guys' (pp. 64–65). If you can't seem to find it there, head to Question 3.

When estimating the size of your insect, be sure to measure the length from the front of the head to the end of the abdomen (as shown in the diagram). Don't include any appendages, such as antennae or wings.

Don't make the mistake of assuming that small insects are younger than large ones, as this is not always the case. Insects such as beetles that undergo complete metamorphosis (see Chapter 2, p. 23) look nothing like their parents. If you find a tiny beetle, it is not a juvenile (young beetles look like grubs), but rather an adult belonging to a species that is small in size.

Question 3. Does it have wings?
Not all insects have wings. Some insects, like silverfish, never grow wings. Insects that swim or burrow under the ground often lose their wings, or never develop wings, to make locomotion easier. The young stages of all insects also lack fully formed wings.

Wings always attach to the thorax

If your insect has wings
   Go to Question 4

If your insect has no wings
   Go to Question 21

⚠ Despite having two pairs of wings, beetles, earwigs and some true bugs have fused their fore wings into a hard shell-like cover, giving them the appearance of having no wings. This makes Question 3 very difficult to answer correctly. Don't despair – the identification key has been designed to correctly identify these insects, no matter what answer you choose for the above question.

⚠ Insects that undergo incomplete metamorphosis (see Chapter 2, p. 22) grow their wings in little sheaths called wing buds, on the outside of their thorax. It can be hard to tell the difference between a fully formed wing and the wing bud of a juvenile insect. However, this identification key will work for either situation, so if you see any evidence of wings, no matter how small, go to Question 4.

💡 Remember – wings always join to the thorax. Anything attached to the head or abdomen is not a wing.

 With the exception of insects that have wing buds (see the warning tip above), if you have answered 'Yes' to Question 3, it means your insect is an adult. Only adult insects have wings, although they may be reduced or absent in some species.

Question 4. How many wings does your insect have?
Most insects have four wings. However, flies only ever have two wings, as their hind wings are reduced to a pair of tiny club-shaped knobs called halteres.

If your insect has two wings
  Answer! Flies, Order Diptera. Go to Chapter 6, p. 225

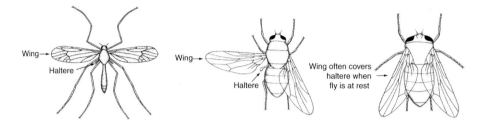

If your insect has four wings
  Go to Question 5

⚠ Many insects keep their delicate hind wings tucked underneath their fore wings, making them difficult to see. Other species have reduced or lost one pair of their wings. Look at the illustrations. If your insect does not look like a fly, then go to Question 5, regardless of how many wings you can count.

 A good way to confirm if your insect is a fly is by finding the little halteres. They are located directly under the fore wings (see illustrations).

⚠ Bees and wasps often hook their wings together for flight, making it look as though they only have two wings when they actually have four. How can you be sure if you have a fly, or a bee or wasp with its wings joined together? Have a look at its abdomen. Bees and wasps have a narrow 'waist' where their abdomen joins their thorax (see the first illustration in Question 20), although it can be hard to see on the furry bodies of bees. Flies don't have a waist and instead have a uniformly broad abdomen. Look at the illustrations of these insects in Table 2.6 to help you distinguish the difference.

💡 Did you know? Mosquitoes are a type of fly – they have only two wings.

✋ Hover flies (see Chapter 4, p. 133) are harmless flies, but some species mimic the bright colouration of stinging bees and wasps to deter predators. Like bees and wasps, hover flies visit flowers to collect nectar and telling them apart can be very difficult. Read the tips above to help you distinguish hover flies from bees and wasps.

Question 5. Do the wings form a hardened shell-like cover?
Beetles and some true bugs have hardened fore wings that form a shell-like cover that protects their delicate hind wings and abdomens.

*If your insect's fore wings form a shell-like cover (which can be hard and shiny, or soft and leathery) that sits **tight** and **flat** over the full length of its abdomen*
Go to Question 6

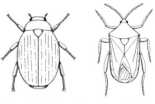
Wings sitting tight and flat over the abdomen

*If your insect's wings are held like a roof over the abdomen, loosely cover the abdomen, sit out away from the body **or** do not reach to the end of the abdomen*
Go to Question 7

Like a roof over the abdomen

Loosely cover abdomen

Wings

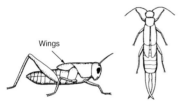
Wings do not reach end of abdomen

Wings sit out away from body

 For insects where the fore wings form shell-like covers:
- the fore wings are not see-through
- the hind wings are usually not visible at all
- the wings extend all the way to the end of the abdomen, but not past it.

For insects where the fore wings **do not** form a shell-like cover, you will often see part of the hind wing poking out from under the fore wings.

There are a few groups of beetles (Order Coleoptera) whose fore wings, while forming a shell-like cover, **do not** extend down the full length of their abdomens. Some longicorn beetles mimic wasps (see p. 193) and have very short wing covers (elytra) so that their membranous hind wings poke out from the bottom (resembling the transparent wings of a wasp). Rove beetles also have cropped wing covers that extend only a short way down their abdomen.

Question 6. How do the wings join together?

*If your insect's wings join together in a straight line that runs the **full length** of the abdomen*
Answer! Beetles, Order Coleoptera. Go to Chapter 6, p. 186

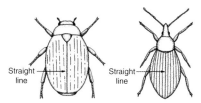

If your insect's wings overlap (resulting in a straight line that extends only part of the way down the abdomen) **or** *if there is no straight line at all*
Answer! Stink bugs, water bugs, plant bugs and their allies, Order Hemiptera, Suborder Heteroptera. Go to Chapter 6, p. 292

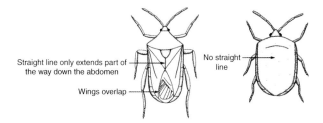

Question 7. Are there pincer-like forceps?

If your insect has a large pair of hard pincer-like forceps at the end of the abdomen
 Answer! Earwigs, Order Dermaptera. Go to Chapter 6, p. 219

If there are no hard pincer-like forceps at the end of your insect's abdomen
 Go to Question 8

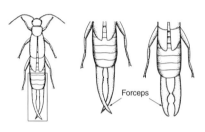

Some insects have a pair of finger-like, thread-like or leaf-like structures called cerci at the end of their abdomen (Table 2.1). These are soft and flexible and very different from the hard forceps of earwigs. If your insect has these cerci rather than pincer-like forceps, go to Question 8.

Adult earwigs may or may not have wings, which is why they pop up again later in the identification key in the section for insects that are wingless.

Diplura (Table 2.7) are tiny, cryptic, soil-dwelling insects so rarely encountered that they are not covered in detail in this book. However, there are a few unusual types that are large (up to 50 mm long) and with pincer-like forceps very similar to earwigs. However, while earwigs may have wings, Diplura are always wingless.

It can be hard to tell the difference between the fully formed wing of an adult earwig and the wing bud (developing wing) of a juvenile, as both are relatively short. However, young earwigs are paler and have shorter antennae than adults.

Question 8. Are the wings and body scaly?
Butterflies and moths have wings and bodies covered in tiny overlapping scales. These scales often give the body and wings a furry or velvety appearance. When scales are present, the wings are generally not transparent (see-through) and instead have solid colours or patterns. Be careful – there are a few exceptions (see below).

If your insect has scales covering its wings and body
    Answer! Butterflies and moths, Order Lepidoptera. Go to Chapter 6, p. 197

If your insect's wings and body are not covered in scales
    Go to Question 9

The scaly body and wings of a silk moth (*Bombyx mori*).

Close-up of the wing of a wanderer (or monarch) butterfly (*Danaus plexippus*) showing the wing scales. Source: J. Dorey.

This variable tigertail dragonfly (*Eusynthemis aurolineata*) does not have a scaly body or wings.

 If you are identifying a dead insect, gently run your finger over the wing or body. If powder comes off on your fingers, it means that body part is covered in scales.

The bee hawk moth (see p. 204) is an unusual type of moth that has clear wings, making it look very much like a bee. Clearwing swallowtails and glasswings are types of butterflies that also have predominantly clear wings. Despite the lack of scales on their wings, all these insects have velvety bodies covered in scales, which is consistent with other butterflies and moths.

38 MORPHOLOGY – WHAT INSECTS LOOK LIKE

Question 9. Are the hind legs used for jumping?
Grasshoppers and crickets stand out from other insects by having chunky thighs on their hind legs, looking a lot like chicken drumsticks! These bulky limbs contain strong muscles that are used for jumping.

If your insect's hind legs are much fatter and quite different looking from its other pairs of legs
Answer! Grasshoppers and crickets, Order Orthoptera. Go to Chapter 6, p. 233

If your insect's hind legs look similar to the mid legs
Go to Question 10

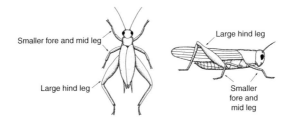

 Sometimes the hind legs of insects are slightly larger than the other legs (e.g. see brown bean bug, p. 123). But if they are all roughly the same shape and design, go to Question 10.

 Some insects jump, but **do not** have enlarged hind legs. If you see your insect jump but can't see big hind legs, go to Question 10.

 Can't really see the legs? Grasshoppers and crickets hold those big hind legs alongside their body, making them very easy to spot. If the legs of your insect are hidden underneath its body, then head to Question 10.

Mole crickets (see p. 89) have enlarged fore legs for digging, and predatory katydids (see p. 240) have big spines on their fore and mid legs. Despite these modifications, they both have large **hind** legs, so belong with other grasshoppers and crickets.

Stick katydids belong to the Order Orthoptera along with other crickets and grasshoppers, but they do not have enlarged hind legs, despite being able to jump. They are restricted to Western Australia and are very uncommon.

Some adult grasshoppers and crickets have wings that are reduced in size. It can be hard to tell the difference between the wings of one of these short-winged adults and the wing buds (developing wings) of a juvenile. In general, adults have longer antennae and more defined genitalia and are larger in size than juveniles, but this can be hard to determine unless you have a juvenile for comparison.

Question 10. Spiky enlarged fore legs for capturing food?
Praying mantids have enlarged fore legs lined with sharp spines. These strong legs snap closed over their prey, squashing them on these spines – a bit like a Venus fly-trap!

If your insect has large fore legs lined with sharp spines
Answer! Praying mantids, Order Mantodea. Go to Chapter 6, p. 250

If your insect's fore legs look the same as the mid legs
Go to Question 11

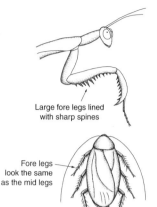

Large fore legs lined with sharp spines

Fore legs look the same as the mid legs

 Some predatory katydids (see p. 240) have big spines on their fore and mid legs. But they also have large **hind** legs for jumping, so belong with the grasshoppers and crickets (see Question 9).

 Mantis flies or *mantispids* (see Chapter 6, pp. 252–253) are a group of lacewings with enlarged fore legs for capturing prey. To tell them apart from true praying mantids, look at their wings. Mantispids have see-through wings with lots of tiny veins, usually held roof-like over their body (see the first two illustrations in Question 12). Praying mantids have leathery opaque fore wings, not as many veins and usually hold their wings flat over their abdomens.

Some adult praying mantids have wings that are reduced in size. It can be hard to tell the difference between the wings of one of these short-winged adults and the wing buds (developing wings) of a juvenile. In general, adults have longer antennae and more defined genitalia and are larger in size than juveniles, but this can be hard to determine unless you have a juvenile for comparison.

Question 11. Long or short?
Stick and leaf insects have elongate bodies that aid in their camouflage and long limbs to help them climb.

*If your insect's body **and** legs are very long and slender and the insect resembles a leaf or a stick*
Answer! Stick and leaf insects, Order Phasmatodea.
Go to Chapter 6, p. 259

If your insect does not fit this description
Go to Question 12

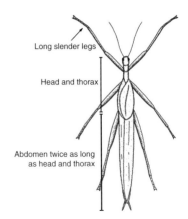

Spiny leaf insects are stick insects that have a habit of flipping their abdomen up over the rest of their body when disturbed (see p. 251). This makes them look a lot shorter than they actually are. However, their leaf-like appearance and their long limbs are characteristic of other stick insects.

Some insects such as dragonflies have very long bodies, but their legs are not long. If you have an insect with a long body but short legs, go to Question 12.

How long is long? Stick insects can reach up to 300 mm in length! But some (especially young ones) are much shorter than this. Look for abdomens that are around twice as long as the rest of the body (head and thorax). Remember – the abdomen begins just behind the spot where the last pair of legs join the body (see illustration above).

Some adult stick insects have wings that are reduced in size. It can be hard to tell the difference between the wings of one of these short-winged adults and the wing buds (developing wings) of a juvenile. In general, adults have longer antennae and more defined genitalia and are larger in size than juveniles, but this can be hard to determine unless you have a juvenile for comparison.

Question 12. Do the wings form a little roof over the body?
Insects can hold their wings out away from their body, can fold them flat, or can rest them like a little roof or an upside-down 'V' over their abdomen.

If your insect's wings are held like a roof over its body
   Go to Question 13

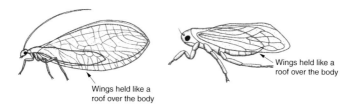

Wings held like a roof over the body

Wings held like a roof over the body

*If your insect's wings fold loosely over the body, sit alongside **or** out away from the body, **or** are held up vertically above the body*
   Go to Question 14

Wings folded loosely over body

Wings alongside body

Wings held away from body

Wings held up vertically above the body

Question 13. How long are the antennae?

The length, shape and structure of antennae can be important characters in identifying insects.

If your insect's antennae are easy to see and are either long and thin or clubbed (gradually thickened at the tip farthest from the head)
Answer! Lacewings, Order Neuroptera. Go to Chapter 6, p. 241

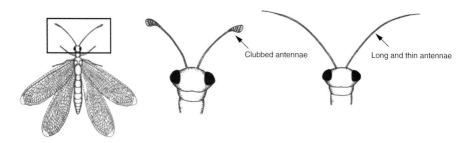

If your insect's antennae are bristle-like (look like tiny hairs), or you can't see them at all
Answer! Cicadas, leafhoppers, planthoppers and their allies, Order Hemiptera, Suborder Auchenorrhyncha. Go to Chapter 6, p. 285

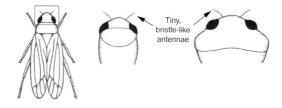

> (?) Can't see your insect's antennae? Sometimes the tiny, bristle-like antennae of cicadas and leafhoppers (like those of some flies) are too small for us to notice. If you can't see any antennae and you have reached this point in the key, it is likely your insect belongs to the Order Hemiptera.

With the exception of a few obscure species, all cicadas, leafhoppers and planthoppers have well-developed wings that extend to the end of their abdomens (and sometimes past it). If your insect has short, flap-like wings that extend only a short way down their back, these are the developing wings (wing buds) of juveniles.

Question 14. What shape are the wings?

Certain insect Orders have characteristic wing shapes and special networks of veins on the wings.

If your insect's wings are long and blade-like with numerous tiny dark veins over the entire wing and the wings are held out away from the body or vertically above the body
   Go to Question 15

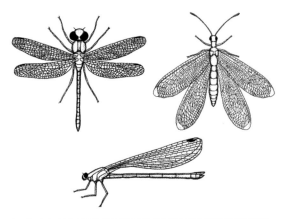

Insects with obvious long blade-like wings with numerous tiny dark veins over the entire wing, and the wings are held out away from the body or held up vertically above the body

If your insect's wings are not long and blade-like, do not have tiny dark veins over the entire wing and are folded flat over the body or parallel to the body
   Go to Question 16

Wings folded flat over body   Wings parallel to body

 Some wasps will hold their wings up vertically above their body. But their wings have far fewer veins, so head to Question 16.

Question 15. How long are the antennae?
The length, shape and structure of antennae can be important characters in identifying insects.

If your insect's antennae are easy to see and are either long and thin or clubbed (gradually thickened at the tip farthest from the head)
Answer! Lacewings, Order Neuroptera. Go to Chapter 6, p. 241

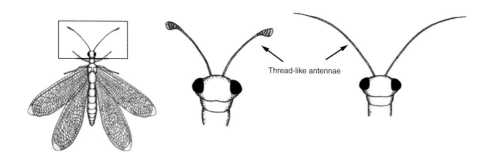

If your insect's antennae are bristle-like (look like tiny hairs), or you can't see them at all
Answer! Dragonflies and damselflies, Order Odonata. Go to Chapter 6, p. 212

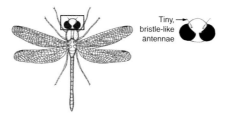

 Can't see your insect's antennae? Sometimes the tiny, bristle-like antennae of dragonflies and damselflies are too small for us to notice. If you can't see any antennae, it is likely your insect belongs to the Order Odonata.

Question 16. How are the wings folded?
Insects that do fold their wings when at rest, may fold them in a variety of ways and this is a feature that can be used to identify them.

If your insect's wings are held alongside the body when at rest
Answer! Bees, wasps, ants and sawflies, Order Hymenoptera. Go to Chapter 6, p. 162

Wings held alongside body

If your insect's wings are folded flat over the abdomen
Go to Question 17

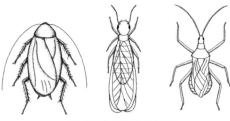
Wings folded flat over abdomen

Question 17. Is the insect small and ant-like?

*If your insect is small (5–35 mm) **and** resembles an ant or termite*
  Go to Question 18

*If your insect is longer than 35 mm **or** does not look like the pictures in this question*
  Go to Question 19

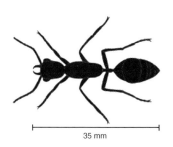
Is your insect smaller than this?

Winged termite

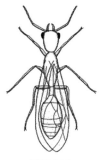
Winged ant

⚠️ When estimating the size of your insect, be sure to measure the length from the front of the head to the end of the abdomen. Do not include any appendages, such as antennae or wings.

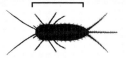

⚠️ Some insects (and even spiders) mimic ants, so look very carefully. See pp. 179–180 (Chapter 6) for examples of ant-mimicking insects.

Question 18. Ant or termite?

If your insect has a wide neck and waist
    Answer! Termites, Order Isoptera. Go to Chapter 6, p. 265

If your insect has a narrow neck and waist
    Answer! Ants, Order Hymenoptera. Go to Chapter 6, p. 174

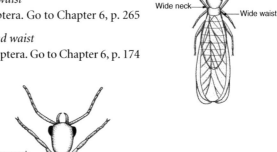

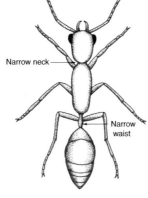

> **?** The ant in the illustration for this question has had its wings removed, to make it easier for you to see the narrow 'waist'. Adult ants and termites may or may not have wings, depending on their job in the colony in which they live. For this reason, they also pop up later in the identification key, in the section that deals with wingless insects.

> **?** 'Neck' and 'waist' are very human-sounding characteristics! To put it into an insect context, the 'neck' is the join between the head and thorax and the 'waist' is the join between the thorax and abdomen.

> 💡 If you are able to see them, the antennae may be used to separate termites from ants. Termites have straight antennae whereas ants have 'elbowed' antennae – that is, a little bend in them, like an elbow (see Table 2.2).

> 💡 Termites and ants may also be distinguished from one another by the colour of their bodies. Termites have abdomens that are usually whitish or cream in colour (sometimes their heads are a slightly darker brown). Ants have abdomens that are seldom whitish or cream in colour.

Question 19. Is it a cockroach?

If your insect has an oval-shaped body, long spiky legs and is brown or black
   Answer! Cockroaches, Order Blattodea. Go to Chapter 6, p. 207

If your insect does not fit this description
   Go to Question 20

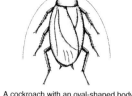
A cockroach with an oval-shaped body and long spiky legs

Colour can be a tricky factor in identifying insects. Some species of cockroaches can be blue, green and other bright colours. But if their body is oval in shape and their legs are spiky-looking, it is a cockroach.

Some adult cockroaches have wings that are reduced in size. It can be hard to tell the difference between the wings of one of these short-winged adults and the wing buds (developing wings) of a juvenile. Adult cockroaches may have bodies that are different in colour or texture from those of juveniles, but this can be hard to determine unless you have a juvenile for comparison.

Question 20. Wasp or true bug?

If your insect has a narrow waist
  Answer! Wasps, Order Hymenoptera. Go to Chapter 6, p. 169

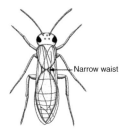

If your insect does not have a narrow waist
  Answer! Stink bugs, water bugs, plant bugs and their allies, Order Hemiptera, Suborder Heteroptera. Go to Chapter 6, p. 292

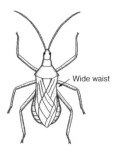

(?) 'Waist' is a very human-sounding characteristic! To put it into an insect context, the 'waist' is the join between the thorax and the abdomen.

Some true bugs have wings that are reduced in size. It can be hard to tell the difference between the wings of one of these short-winged adults and the wing buds (developing wings) of a juvenile. In general, adults have longer antennae and more defined genitalia and are larger in size than juveniles, but this can be hard to determine unless you have a juvenile for comparison.

Question 21. Is it grub-like?
   While grubs, maggots and caterpillars each have unique body shapes (see Chapter 3, pp. 68–70), their overall appearance is very different from that of other wingless insects.

If your insect looks like a maggot, grub or caterpillar
   Go to 'Maggots, grubs and caterpillars', p. 64

If your insect does not resemble the illustrations in this question
   Go to Question 22

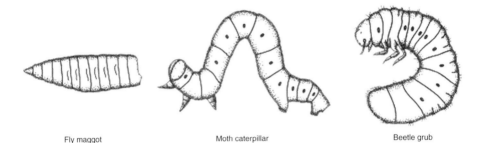

Fly maggot   Moth caterpillar   Beetle grub

If your insect does not resemble these illustrations, head to pp. 68–70 and look at the other examples of insect larvae. If you can't find it there and your insect has long legs and a hard-looking exoskeleton, go to Question 22.

Insects that look like maggots, grubs or caterpillars are the young or *larval* stages of insects that undergo complete metamorphosis (see Chapter 2, p. 23). However, there are some exceptions (see the tip below).

Occasionally, adult insects are *larviform* – that is, they resemble maggots, grubs and caterpillars (insect larvae). For example, queen termites (see p. 269) have enormous, grub-like abdomens filled with thousands of eggs. Female scale insects (p. 152) and mealybugs (see pp. 152–153) are sedentary, wingless and often legless, superficially resembling maggots. The females of some obscure beetle species are almost indistinguishable from their grub-like larvae, and some species of moths have fat, grub-like females that lack wings.

Question 22. Does it have tails at the end of its abdomen?

If your insect has three long, thread-like tails at the end of its abdomen
    Answer! Silverfish, Order Thysanura. Go to Chapter 6, p. 256

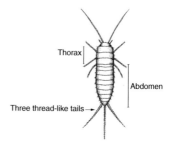

If your insect does not have thread-like tails at the end of the abdomen or if it has fewer than three
    Go to Question 23

⚠ Don't confuse tails with legs. Remember, legs always join to the thorax. Tails will always attach at the end of the abdomen. Use the illustration above to help you.

🐝 Silverfish remain wingless throughout their lives: it is difficult to tell juveniles apart from adults as both lack wings. Silverfish have a unique way of growing (see Chapter 2, p. 22) and continue to moult even as adults.

Question 23. Are there pincer-like forceps?

If your insect has a large pair of hard pincer-like forceps at the end of the abdomen
 Answer! Earwigs, Order Dermaptera. Go to Chapter 6, p. 219

If there are no hard pincer-like forceps at the end of your insect's abdomen
 Go to Question 24

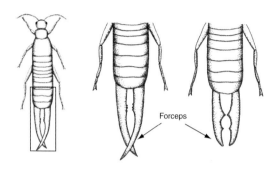

Forceps

⚠ Some insects have a pair of finger-like, thread-like or leaf-like structures called cerci at the end of their abdomen (Table 2.1). These are soft and flexible and very different from the hard forceps of earwigs. If your insect has these cerci rather than pincer-like forceps, go to Question 24.

 Adult earwigs may or may not have wings. Juvenile earwigs can be distinguished from wingless adults by their shorter antennae and paler colour.

✋ Diplura (Table 2.7) are tiny, cryptic, soil-dwelling insects so rarely encountered that they are not covered in detail in this book. However, there are a few unusual types that are large (up to 50 mm long) and with pincer-like forceps very similar to earwigs. To tell them apart – earwigs have bodies **and** forceps that are dark in colour and they have visible eyes. Diplura have pale bodies with dark forceps and never have eyes.

54 MORPHOLOGY – WHAT INSECTS LOOK LIKE

Question 24. Are the hind legs used for jumping?
  Grasshoppers and crickets have chunky thighs on their hind legs, containing strong muscles for jumping. They look a bit like chicken drumsticks!

If your insect's hind legs are much fatter and quite different-looking from the mid legs
  Answer! Grasshoppers and crickets, Order Orthoptera. Go to Chapter 6, p. 233

If your insect's hind legs look similar to the mid legs
  Go to Question 25

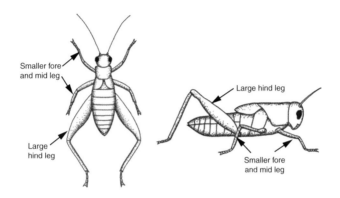

 Sometimes the hind legs of insects are slightly larger than the other legs (e.g. see brown bean bug, p. 123). But if they are all roughly the same shape and design, go to Question 25.

Mole crickets (see p. 89) have enlarged fore legs for digging, and predatory katydids (see p. 240) have big spines on their fore and mid legs. But despite these modifications, they both have large **hind** legs, so belong with other grasshoppers and crickets.

Stick katydids belong to the Order Orthoptera along with other crickets and grasshoppers, but they do not have enlarged hind legs, despite being able to jump. They are restricted to Western Australia and are very uncommon.

MORPHOLOGY – WHAT INSECTS LOOK LIKE   55

 Can't really see the legs? Grasshoppers and crickets hold those big hind legs alongside their body, making them very easy to spot. If the legs of your insect are hidden underneath its body, then head to Question 25.

 Some insects jump, but **do not** have enlarged hind legs. If you see your insect jump but can't see big hind legs, go to Question 25.

Some adult grasshoppers and crickets are wingless, making them difficult to distinguish from juveniles (which also lack wings). In general, juveniles have shorter antennae than adults, but this can be hard to determine unless you have an adult for comparison.

Question 25. Spiky enlarged fore legs for capturing food?
Praying mantids have large fore legs lined with sharp spines. Their prey is squashed on the spines and immobilised.

If your insect has large fore legs lined with sharp spines
    Answer! Praying mantids, Order Mantodea. Go to Chapter 6, p. 250

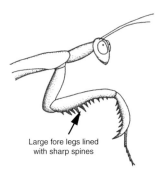

Large fore legs lined with sharp spines

If the fore legs of your insect look the same as the mid legs
    Go to Question 26

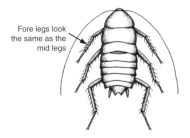

Fore legs look the same as the mid legs

 Some predatory katydids (see p. 240) have big spines on their fore and mid legs. But they also have large **hind** legs for jumping, so belong with the grasshoppers and crickets (see Question 24).

Some adult praying mantids are wingless, making them difficult to distinguish from juveniles (which also lack wings). In general, juveniles have shorter antennae than adults, but this can be hard to determine unless you have an adult for comparison.

Question 26. Long or short?
Stick and leaf insects have elongate bodies that aid in their camouflage and long slender legs used for climbing.

*If your insect's body **and** legs are very long and slender and it resembles a leaf or a stick*
Answer! Stick and leaf insects, Order Phasmatodea. Go to Chapter 6, p. 259

If your insect does not fit this description
Go to Question 27

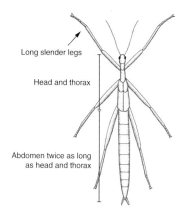

How long is long? You may be identifying a young stick insect, which may be quite small in size compared to the enormous adults. Instead of getting hung up on size, concentrate on the shape of the body and the length of the legs. Look for abdomens that are around twice as long as the rest of the body (head and thorax). Remember – the abdomen begins just behind the spot where the last pair of legs join the body (see illustration above).

Some adult stick and leaf insects are wingless, making them difficult to distinguish from juveniles (which also lack wings). In general, juveniles have shorter antennae than adults, but this can be hard to determine unless you have an adult for comparison.

Remember also that some insects can have long bodies, but their legs will be quite short. If this is the case, go to Question 27.

Question 27. Does it have a narrow neck and waist?

If your insect has a narrow neck and waist
    Answer! Ants, Order Hymenoptera. Go to Chapter 6, p. 162

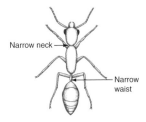

If your insect does not have a narrow neck and waist
    Go to Question 28

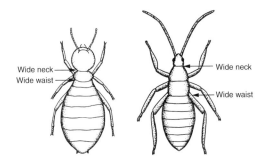

 'Neck' and 'waist' are very human-sounding characteristics! To put it into an insect context, the 'neck' is the join between the head and the thorax and the 'waist' is the join between the thorax and the abdomen.

⚠ The insect **does not** have to look exactly like the pictures in this question. As long as it does not have a narrow neck and waist, head to Question 28.

 If you have identified your insect as an ant, then it is definitely an adult. Ants undergo complete metamorphosis (Chapter 2, p. 23) – a lifecycle similar to that of a butterfly. Rather than looking like tiny ants, juveniles are maggot-like grubs.

Question 28. Is it a termite?

If your insect is small (5–15 mm long), with no visible eyes and is pale brown, yellow or cream
  Answer! Termites, Order Isoptera. Go to Chapter 6, p. 265

15 mm

If your insect does not fit this description or does not resemble the pictures in this question
  Go to Question 29

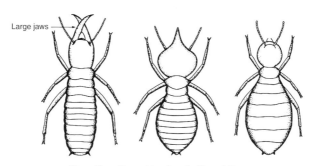

Some different types of termites, all with no visible eyes

⚠ When estimating the size of your insect, be sure to measure the length from the front of the head to the end of the abdomen. Don't include any appendages, such as antennae.

❓ The illustrations show various types of termites which have different shaped heads, based on their role within the termite colony. The first two are soldier termites, a special group of individuals that guard the nest and the queen. The modifications to their heads help them fight against enemies – the first has large jaws for biting and the second has a pointy nozzle on its head for spraying toxic chemicals. The third is a worker termite (which does not defend the nest) and so lacks any such modifications.

⚠ Some soldier termites have large jaws (see the first of the group of three termites) and can be mistaken for lacewing larvae (see Question 29). As a rule, lacewing larvae generally have jaws the same colour as the rest of their bodies. The jaws of soldier termites are usually much darker than the rest of their bodies.

Termites are social insects that live in colonies and within each colony there are several castes – groups of individuals with particular roles. Reproductive termites have wings and are dealt with earlier in the identification key (see Question 18), but all other termites in the colony (both adults and juveniles) are wingless. Soldier termites are wingless adults, with well-developed heads that come in one of two different forms, depending on the species (see illustration). Worker termites are adults that lack wings and any other structural modifications, making them almost impossible to distinguish from the similar-looking nymphs.

Question 29. Is it an antlion?

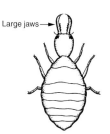

Large jaws

If your insect has a large pair of jaws poking out from its head
Answer! Antlions and lacewing larvae, Order Neuroptera. Go to Chapter 6, p. 241

If your insect does not have a large pair of jaws poking out from its head
Go to Question 30

Antlions are the juvenile stages or *larvae* of lacewings (winged insects that resemble dragonflies). Like butterflies, lacewings undergo complete metamorphosis (Chapter 2, p. 23) and start life looking very different from their winged parents.

⚠ Some soldier termites have large jaws (see the first of the group of three termites in Question 28) and can be mistaken for antlions and other lacewing larvae. As a rule, lacewing larvae generally have jaws the same colour as the rest of their bodies. The jaws of soldier termites are usually much darker than the rest of their bodies.

62 MORPHOLOGY – WHAT INSECTS LOOK LIKE

Question 30. Is it a cockroach?

If your insect is oval-shaped, brown, black or grey, with spiky legs and long thread-like antennae
   Answer! Cockroaches, Order Blattodea. Go to Chapter 6, p. 207

If your insect does not fit this description
   Go to Question 31

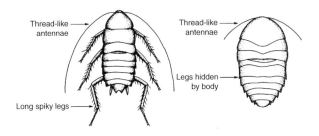

> Colour can be a tricky factor in identifying insects. Some species of cockroaches can be blue, green and other bright colours. But if their body is oval-shaped and their legs are spiky-looking, it is a cockroach.

> Some cockroaches (like the picture in this question) have wide bodies that cover up their spiky legs. But they will be oval-shaped, have long thread-like antennae and are usually brown, so concentrate on these characters instead.

> Some adult cockroaches are wingless, making them difficult to distinguish from juveniles (which also lack wings). Young cockroaches may have bodies that are different in colour and texture from those of the adults, but this can be hard to notice unless you have an adult for comparison.

Question 31. Beetle or true bug?

If you have not managed to place your insect into any of the other Orders by now, it will most likely fall into one of the two below.

*If your insect has a hard, often shiny shell-like cover, with one or more straight lines running **vertically** down the **full length** of the abdomen*

Answer! Beetles, Order Coleoptera. Go to Chapter 6, p. 186

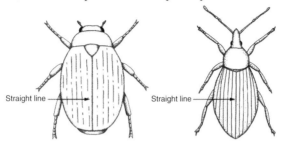

*If your insect has a leathery shell-like cover, with **no** straight lines running vertically down the length of the body, **or** it has a straight line that reaches only part of the way down the abdomen, **or** horizontal lines running across the abdomen*

Answer! Stink bugs, water bugs, plant bugs and their allies, Order Hemiptera, Suborder Heteroptera. Go to Chapter 6, p. 292

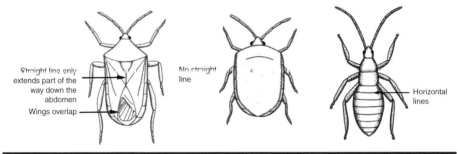

With the exception of the true bug nymph in the second illustration in this question (with the horizontal lines on its abdomen), insects of these types are adults and do have wings, but because they are folded tight and flat over the body you may not have noticed them, which is why you have ended up at this point. Beetles and stink bugs, water bugs, plant bugs and their allies are sometimes very difficult to tell apart, so look at the summary box for both Orders (Chapter 6) to help you make your final decision.

Some adult bugs belonging to the Suborder Heteroptera are wingless, making them difficult to distinguish from juveniles (which also lack wings). In general, juveniles have shorter antennae than adults, but this can be hard to determine unless you have an adult for comparison.

The little guys – insects less than 5 mm long

If you answered 'Yes' to Question 2 in the identification key, the body of your insect is less than 5 mm long. This section will help you to learn more about insects that are tiny in size.

All insects seem small to us, but some types of insects (and some stages of insects) are minute, with bodies that are only a few millimetres long. Some belong to Orders where some species are small and other species are comparatively a lot larger. Others belong to Orders where **all** species are tiny, allowing them to go unnoticed as they exploit habitats and food sources that larger insects cannot. For example, fleas and lice often go undetected on the bodies of other animals, because they are minute.

Figures 3.2–3.7 illustrate different groups of minute insects. You can learn more about the biology of these insects by referring to the information pages for the Order to which they belong.

Maggots, grubs and caterpillars

There are many different maggot, grub and caterpillar-like creatures in our homes and gardens. Some are the juvenile stages of insects that will transform into winged adults through a process called complete metamorphosis (see Chapter 2, p. 23). With practice, you will begin to recognise these various types of larvae and the insects they will turn into as adults. However, insect larvae can easily be confused with millipedes and centipedes – regular visitors to our gardens that are not insects at

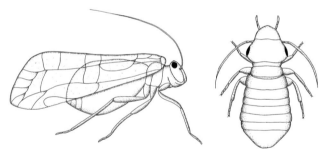

Fig. 3.2: Booklice, Order Psocoptera (p. 194). These are tiny, flesh-coloured insects you may see running across your desk or in your pantry.

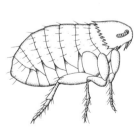

Fig. 3.3: Fleas, Order Siphonaptera (p. 221). These wingless, jumping insects are usually black or brown and are associated with pets and their bedding.

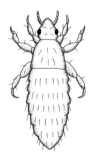

Fig. 3.4: Lice, Order Phthiraptera (p. 246). These wingless, usually brown insects are found on humans, birds and other warm-blooded animals.

MORPHOLOGY – WHAT INSECTS LOOK LIKE 65

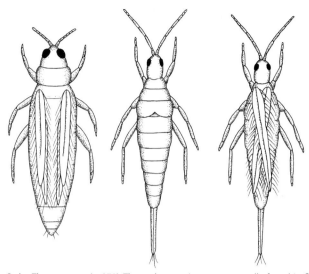

Fig. 3.5: Thrips, Order Thysanoptera (p. 270). These elongate insects are usually found in flowers or on the surface of leaves.

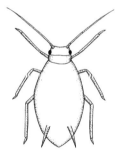

Fig. 3.6: Aphids, Order Hemiptera, Suborder Sternorrhyncha (pp. 106, 278). These fat-bodied insects can be yellow, black, pink or green and are found on the growing shoots and leaves of plants.

Fig. 3.7: Bed bugs, Order Hemiptera, Suborder Heteroptera (pp. 84, 292). These brown insects have rounded bodies and are found in bedding and furniture in dodgy hotels and other poorly maintained accommodation.

> Does your insect not look like any of the illustrations in Figs 3.2–3.7? There are certain types of flies, ants, wasps, beetles and other insects that are minute. However, despite their size, these tiny individuals share the same key features seen in larger members of their Order, so refer back to the identification key and move on to Question 3.

all, but look-alikes belonging to other arthropod groups. This section helps you recognise different types of insect larvae and teaches you how to distinguish them from things that are not insects.

Distinguishing insect larvae from centipedes and millipedes

Some insect larvae have legs; others are legless. At most, insect larvae have six true legs. Some butterfly, moth and sawfly larvae have additional leg-like suction cups along their abdomens known as prolegs, which may be mistaken for true legs. If you can't recognise these false legs, you might assume your insect larva has more than six legs and confuse it with non-insect groups such as centipedes and millipedes. Figure 3.8 helps you distinguish true legs from prolegs and

Insect larva
- at most six legs, all located near head
- legs may be absent
- may have additional suckers (prolegs) on some (but not all) abdominal segments
- body usually cylindrical

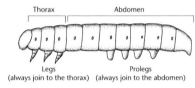

Centipede
- more than six legs
- legs never absent
- legs found along the entire length of the body with one pair of legs on every body segment
- body slightly flattened

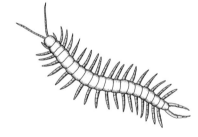

Millipede
- more than six legs
- legs never absent
- legs found along the entire length of the body with two pairs of legs on every body segment
- body cylindrical

Fig. 3.8: Distinguishing insect larvae from centipedes and millipedes.

Aquatic insect larvae

Beware of any maggot, grub or caterpillar-like creatures you find in underwater habitats. Many aquatic insect larvae have long leg-like gills attached to the entire length of their bodies, making them look like millipedes and centipedes. However, millipedes and centipedes do not live in the water. They are found only in terrestrial environments (i.e. on land).

This spiny mayfly nymph (*Coloburiscoides* sp.) has gills along its abdomen which may be mistaken for legs. Source: Gooderham and Tsyrlin (2002).

Legless insect larvae

Some insect larvae are legless, but are classified as insects because they grow six legs when they become adults. Legless organisms can be very difficult to identify. Slugs and worms (not covered in this book) superficially resemble legless insect larvae, such as maggots. If you can't find your critter in Figs 3.9–3.12, it is probably not an insect.

outlines the differences between insect larvae, centipedes and millipedes.

Different types of insect larvae

Once you have confirmed your critter is indeed an insect, you can use this section to compare it to some of the most commonly encountered insect larvae in our homes and gardens. These illustrations focus on basic body shape (e.g. the presence or absence of legs, prolegs and well-developed head capsules) to distinguish between the larvae of different insect Orders.

At first, it may seem as though all insect larvae look the same but, with practice, you will soon be recognising the different Orders with confidence. The classification of insect larvae into more specific types (e.g. Suborders, Families etc.) often comes down to arrangements of tiny hairs on the body, not easily visible to the naked eye. In these cases, a microscope is necessary for accurate identification.

Fly larvae, Order Diptera (see Chapter 6, p. 225)

The larvae of flies are usually called maggots. Their bodies are shaped like a torpedo, tapering gradually towards the head. They have no eyes or legs and their mouthparts are usually not visible. Some aquatic fly larvae such as mosquito wrigglers (see Chapter 4, p. 100) are unusual in this regard. They have a well-developed head and body.

Fly larvae almost always live in moist environments such as soil, mud, water and animal dung, and on decaying plant and animal matter (e.g. compost, rubbish, the bodies of dead animals).

Bee, wasp, ant and sawfly larvae, Order Hymenoptera (see Chapter 6, p. 162)

Hymenoptera larvae vary in appearance and this relates to their type (bee, sawfly etc.) and where they feed. The paper wasp larva (Fig. 3.10) illustrates the typical legless maggot-like bodies of bee, wasp and ant larvae. However, their strong head capsules and distinctive shape help distinguish such larvae from the larvae of flies (Fig. 3.9). Hymenoptera larvae occupy a wide variety of habitats (e.g. underground, inside the bodies of other insects, within plant galls and in communal nests) and food sources.

The larvae of sawflies (not really a fly, but a type of primitive wasp) have well-developed legs and often a large spike at the end of their bodies (Fig. 3.10). They may also have a series of suction cups along their abdominal segments, called prolegs, to help them grip onto plants. Sawfly larvae feed on

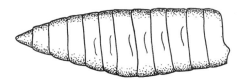
Blowfly larva

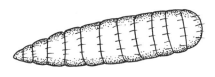
Soldier fly larva

Fig. 3.9: Commonly encountered fly larvae.

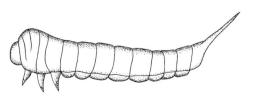
'Spitfire caterpillar' (sawfly larva)

Paper wasp larva

Fig. 3.10: Commonly encountered wasp larvae.

leaves (commonly gum leaves) and often feed in groups (see Chapter 4, p. 127).

Beetle larvae, Order Coleoptera (see Chapter 6, p. 186)

Beetle larvae are highly variable in appearance (Fig. 3.11) and we usually refer to them as 'grubs'. Generally they have a well-developed head capsule, strong jaws and six legs. Some beetle larvae are similar in appearance to sawfly larvae (Fig. 3.10), but are usually a lot hairier.

Curl grubs are the larvae of scarab beetles (a group that includes rhinoceros, dung, cane and Christmas beetles). They have thick bodies curled into a distinct 'C' shape. Other beetle larvae, such as those of leaf beetles, ladybird beetles and click beetles, have more slender bodies and legs. Beetle grubs can be found in a wide variety of habitats, including soil, compost, inside living and dead timber, underwater and on the surface of leaves.

Butterfly and moth larvae, Order Lepidoptera (see Chapter 6, p. 197)

Butterfly and moth larvae (known as caterpillars) have large head capsules, strong chewing mandibles and well-developed legs (Fig. 3.12). The bodies of caterpillars tend to be longer and more cylindrical than those of other insect larvae. They may be smooth in appearance (e.g. hawk moth larvae, inch worms, many butterfly larvae). 'Hairy caterpillar' is a term that covers several species whose larvae are clothed in long, itchy hairs. Some (e.g. cup moth larvae) are covered in sharp spines, which can deliver a painful sting on contact.

Butterfly and moth caterpillars often have prolegs along their abdomen to help them cling to the plants on which they feed. To the naked eye, it is very difficult to distinguish caterpillars from the larvae of sawflies (Fig. 3.10).

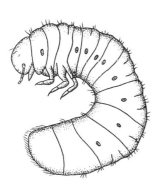

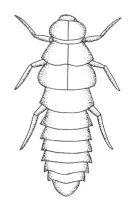

'Curl grub' (Rhinoceros beetle larva)   Click beetle larva   Ladybird beetle larva

Fig. 3.11: Commonly encountered beetle larvae.

70 MORPHOLOGY – WHAT INSECTS LOOK LIKE

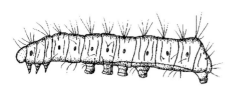

'Hairy caterpillar'
(noctuid moth larva)

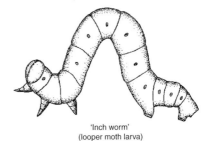

'Inch worm'
(looper moth larva)

Cup moth larva

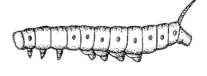

Hawk moth larva

Fig. 3.12: Commonly encountered moth larvae.

4
Habitat – where insects live and occur

When it comes to the social lives of insects and humans, we have more in common than you may think. We both tend to have regular hangouts where we gather to eat our favourite foods, socialise and even search for potential partners. Some insect patrons are so loyal to a particular place, they can be identified on this factor alone. This chapter helps you to identify an insect based on its habitat, rather than on its appearance.

Using this chapter

This chapter provides information pages for specific habitats around your home and garden where insects commonly occur. From the options provided in Fig. 4.1, select the habitat in which you found your insect. Go to the section for that particular habitat, where you can learn more about some of the different insects commonly found there.

The Order, common and scientific name (where possible) is provided for each insect listed in a given habitat, along with one or more photos. You can look up the Order that your particular insect belongs to in Chapter 6 to learn more about its lifecycle, biology and survival tactics. In instances where the insects listed for a particular habitat belong to multiple Genera (e.g. the many types of weevils that can be found on trees and shrubs) the name of the Family to which the insects belong is also included, which is handy for internet searches and further research. Be sure to also check out any further reading suggestions for specific habitats in the Further Reading list at the back of the book.

Some insects are observed in multiple habitats. As you read through this chapter, an insect you would expect to feature in a given section may not be there, and you are instead referred to another part of the book. This is because they are listed under a habitat where they are more frequently observed, or they are described on their Order information page in Chapter 6.

For example, you may notice Christmas beetles feeding on native gum leaves in your garden and, wishing to learn more about them, refer to the native trees and shrubs section (p. 125). Christmas beetles, however, do not feature on that page and you are instead referred to 'Around lights' (p. 86). It is my experience that Christmas beetles are most often seen by people as they buzz around artificial lights, and so I have chosen to list them under this alternative (more likely) habitat.

Another example is that you may see butterflies visiting the flowers in your garden, but on the 'In and around flowers' section you are directed straight to the information page for their Order, Lepidoptera. This is because, with a few very rare exceptions, **all** butterflies drink nectar from flowers and so everything you need to know about butterflies of all kinds is covered on their Order information page.

72 HABITAT – WHERE INSECTS LIVE AND OCCUR

Fig. 4.1: Choose the habitat where you found your insect. (1) Kitchen and pantry, p. 74. (2) Living room, p. 78. (3) Bathroom, p. 83. (4) Bedroom, p. 84. (5) Around lights, p. 86. (6) In water, p. 93. (7) Soil, leaf litter and compost, p. 102. (8) On trees and shrubs, p. 106. (9) On citrus trees, p. 116. (10) In the vegetable garden, p. 120. (11) On native trees and shrubs, p. 125. (12) In and around the lawn, p. 128. (13) In and around flowers, p. 132. (14) On the bodies of animals – the bloodsuckers, p. 135. (15) In large groups – masses and migrations, p. 138.

Tips for getting started

Before you get started, here are some tips that you need to consider as you navigate through this chapter.

Tip 1. Tenant or traveller?

Let's say you find an insect on a rose bush. Does the insect actually live on the plant, or use it for food? Perhaps the insect is just sunning itself, has been blown there by the wind or is simply resting as it passes through the area.

Try to observe your insect for a while before you rush to any conclusions. Can you see it actively feeding? Are there any signs of feeding damage, such as bite marks on the leaves? Is the insect with other individuals of the same kind? If you answer 'Yes' to any of these questions, it is likely that the insect is in its habitat and you can proceed through this chapter. If not, refer to Chapter 3 where you can focus your search using the insect's appearance, rather than on where you found it.

Tip 2. Unfussy eaters

Many herbivorous insects restrict their diet to a single food source. For example, the caterpillars of some butterflies only eat leaves from a single species of plant. However, some plant-feeding insects sit at the opposite end of this spectrum and munch on a variety of vegetation, including native trees, grasses, ornamental shrubs, fruit and vegetables.

Assigning insects with broad diets to a particular habitat (e.g. on native trees and shrubs) is not straightforward, so it is wise to flip through the different vegetation types in this chapter (using Fig. 4.1 as a guide), when identifying insects you have found on plants.

Tip 3. Making the cut
The insects featured in this chapter are relatively common, with a wide distribution within Australia, and may well turn up in your home or garden on a regular basis. Several small, obscure insect Orders have been omitted from this book and the insect you are trying to identify may belong to one such Order. Refer to Table 2.7 (pp. 19–20), which provides a brief description of the insect Orders omitted from this book and the habitats they usually occupy. If your mystery insect lives in the same habitat as one of these Orders, you have a starting point for further research.

Tip 4. Can't find it here?
It is not possible to include every insect that may occur in a given habitat. At the end of each habitat listed in this chapter, you'll find a list of other insects that are also commonly found in that particular habitat. You will be referred to pages in this book where you can learn more about them.

Still can't find it? Try using the identification key in the morphology chapter (Chapter 3) to see if you can identify your insect by its appearance rather than by means of its habitat. Or refer to Chapter 5, 'Clever clues', to see if you can find it there.

Kitchen and pantry

Many insects enjoy eating the same foods as we do. Flour, pasta, dried fruit, nuts and even chocolate are readily consumed by certain species. Some insects squeeze their way into cracks in our cupboards and deposit their eggs in and around foodstuffs. The larvae of some pantry moths can chew through plastic bags or crawl under screw-top lids. Other insects infest products such as flour and cereals while they are in storage, so that when we purchase food from the supermarket we inadvertently bring them home with us, and the longer we store the food the more insects there will be.

Cigarette beetle

Lasioderma serricorne

Order Coleoptera

Many species of small brown beetles infest stored products within our homes. The cigarette beetle, as the name implies, can be found in tobacco products and happily drills numerous holes through cigarettes and cigars. These beetles and their larvae are very secretive, shunning daylight and sheltering among their food of choice, which can also extend to nuts, herbs and spices, dried pet food and foam packaging material.

Size: Adults 3–4 mm

Goodie or baddie? These beetles infest a variety of stored food products, leading to a lot of wastage.

Even the hottest cayenne pepper may be attacked by cigarette beetles (*Lasioderma serricorne*).

Rice weevil

Sitophilus oryzae
Order Coleoptera

Rice weevils differ from other pantry beetles by the long 'snout' on their heads, a feature that is obvious on most weevils. The adults are dark brown or black and have four pale patches on their backs. They are good climbers, swiftly scaling vertical surfaces such as storage containers and pantry walls. Female beetles bite a small hole in the kernel of grains, deposit an egg, then seal the hole with a waxy plug. They can lay 150–400 eggs throughout their life. The grub-like larvae are tiny and generally remain unnoticed as they chew their way through cereals, rice and other grains.

Size: Adults 3–4 mm

Goodie or baddie? The tiny size of the adults and larvae means they often go unnoticed as they munch their way through our stored food products.

Rice weevils (*Sitophilus oryzae*) infest stored products such as cereals, grains and dried pasta. Source: Science Image.

Indian meal moth
Plodia interpunctella
Order Lepidoptera

The Indian meal moth is a species of 'pantry moth' – a collective term used to describe several different types of moths that invade our stored food products. This particular moth is easily distinguished by its two-toned wings – pale grey towards the head and a dark reddish-brown towards the back edge. As with most moths, they are attracted to lights and can be seen around ceilings and walls at night. During the day, you may see them flying in the pantry if you disturb them. A female moth can lay between 150–400 eggs and these are often brought into your pantry on the outside of packaged food. The larvae of these moths are small and pale and have a habit of draping silk webbing where they have been feeding. They have expensive tastes and love munching their way through chocolate, nuts, spices and dried fruit.

Size: Adult wingspan 6–20 mm; larvae up to 15 mm

Goodie or baddie? This moth is a real pain to eliminate from your pantry and re-infestations are common.

Adult Indian meal moths (*Plodia interpunctella*) on cereal grains. Source: Science Image.

Vinegar fly
Drosophila sp.
Order Diptera

Vinegar flies are also called 'fruit flies', however they are not to be confused with the large yellow and brown flies of the same name (belonging to the Family Tephritidae) that are serious pests in fruit crops. Vinegar flies are tiny yellowish-brown insects with red eyes and clear wings. These flies are found on every continent (except Antarctica) and are spread to new locations by human activities, such as the transportation of infested produce by boats. In our homes they are attracted to the smells of rotting fruit, decomposing compost and fermented liquids such as beer and wine. If you have ever found a tiny insect swimming laps in your glass of chardonnay, it is most likely a vinegar fly. The females deposit their eggs (around 400 of them!) into soft fruit, where their tiny maggots feed and develop. Once they are big enough, the maggots will leave the fruit in search of a dry place to pupate (e.g. the bottom or sides of your fruit bowl). The maggots form tiny brown capsules and inside this, transform into an adult fly.

Size: Adults 2–4 mm

Goodie or baddie? Because vinegar flies grow quickly, lay lots of eggs and are easy to breed and care for, they are studied in biological research laboratories around the world. So the next time you find an annoying cloud of vinegar flies hanging around your fruit bowl, remember that these tiny insects may hold the key to postponing ageing, defying death or curing hereditary diseases.

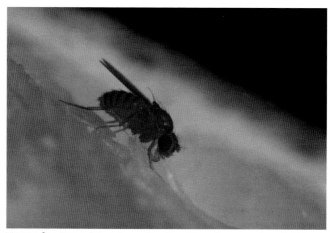

Vinegar flies (*Drosophila* sp.) are often found hovering around our fruit bowls. Source: J. Dorey.

Other insects commonly found in the kitchen and pantry
American cockroach, p. 80
Ants, p. 174
German cockroach, p. 79

Living room

The following insects may be found not only in your living room, but in most areas of your home where there are similar food sources.

Museum beetle

Anthrenus (Florilinus) museorum

Order Coleoptera

Also referred to as carpet beetles, adult museum beetles are small, oval-shaped and have patterned grey, brown and white bodies. Their larvae are brown and hairy-looking and can move quite quickly. Both adults and larvae infest dried animal materials such as woollen carpets and clothes, animal furs and hides, and animal specimens such as stuffed birds and insect collections.

Size: Adults 2–3 mm

Goodie or baddie? These beetles and their larvae can be very destructive and often go unnoticed until they have caused significant damage.

Museum beetles (*Anthrenus (Florilinus) museorum*) cause damage to animal materials, hides and stuffed animal specimens. Source: Science Image.

German cockroach
Blattella germanica
Order Blattodea

The German cockroach is pale yellow-brown, with a dark brown stripe running down each side of its shield-like thorax. Females lay 35–48 eggs at a time, enclosed within a purse-shaped bag called an ootheca (pl. oothecae) (see Chapter 5, p. 159). The females often carry these egg cases around within their bodies, but protruding from the end of their abdomen. A single female can produce up to eight of these oothecae in her life, totalling over 300 individual eggs.

As with other pest cockroaches, German cockroaches eat any organic material in your homes, including foodstuffs, paper, soap, toothpaste, carpet and dried pet food.

Size: Adults up to 15 mm

Goodie or baddie? German cockroaches breed up to enormous populations and you may find their small black droppings scattered around your benches and cupboards.

German cockroaches (*Blattella germanica*) are much smaller than American cockroaches, but are no more welcome in our homes. Source: Sarah Camp, Attribution–NoDerivs 2.0 Generic.

American cockroach

Periplaneta americana
Order Blattodea

American cockroaches are large dark brown insects with pale yellow-brown markings on their thorax. They have long spiky brown legs. The Australian cockroach (*Periplaneta australasiae*) is almost identical, with extra yellow markings on the edges of the wings being the obvious difference between the species. Both are largely nocturnal, hiding in dark places during the day and emerging at night to feed on almost any kind of plant or animal material in your homes. The adults have long wings and can sometimes be seen flying around lights at night. They can also breed up in large numbers in your compost bins.

Size: Adults 35–40 mm

Goodie or baddie? These cockroaches are large and smell disgusting, making them a very unwelcome visitor in your home.

American cockroach (*Periplaneta americana*) scavenging for crumbs.

Flesh flies
Many species
Order Diptera, Family Sarcophagidae

If the name 'flesh flies' makes your skin crawl, fear not! These flies do not harm us in any way, but rather lay their eggs on the decomposing bodies of dead animals. They have large red eyes, three black stripes on their grey thorax and a black and grey 'checkerboard' pattern on the back of their abdomen. These flies sometimes rest on your walls and ceilings, most likely drawn indoors in search of dead rodents in internal cavities within your home.

Size: Adults up to 15 mm

Goodie or baddie? Although their taste in food is questionable, flesh fly maggots help to break down and recycle the carcasses of dead animals.

Adult flesh fly (Family Sarcophagidae) showing the distinctive grey and black stripes on its thorax.

Granny's cloak moth

Speiredonia spectans
Order Lepidoptera

This handsome moth has large brownish-grey wings that reflect a beautiful purple hue in the light. A large eyespot on each wing is used to deter predators such as birds. These moths favour cool, dark places and occupy caves and tree hollows during the day. They also seek shelter in and around our homes and can be found in sheds, garages, basements and cupboards and around the eaves of your roof, sometimes resting together in large groups.

They frequently sit on the walls and ceilings of outdoor toilets, earning them the nickname 'dunny moth'.

Size: Adult wingspan 75 mm

Goodie or baddie? The larvae of this moth feed on the foliage of wattles and so cause no damage to fruit and vegetable crops. Many people freak out when they find groups of these rather large moths in their house or garage, but these moths are harmless.

Granny's cloak moths (*Speiredonia spectans*) can be found resting on the walls and ceilings of your home. Source: J. Dorey.

Other insects commonly found in the living room

Ants, p. 174
Booklice, p. 194
Silverfish, p. 256

Bathroom

While the insect detailed below may frequently be found in our bathroom and toilet, many of the critters from other areas of your home may also venture there in search of moisture. Thirsty ants can be found huddling around droplets of water and cockroaches sneak in for a drink (and perhaps to nibble on your soap or any traces of paste left on your toothbrush)!

Moth flies
Many species
Order Diptera, Psychodidae

Moth flies, with their hairy grey bodies and wings, resemble tiny moths. If you look closely, however, you will see they only have a single pair of wings, a trait that is characteristic of flies. Moth flies are tiny and can be found in damp areas of your home such as the bathroom, toilet and laundry, often clinging to the walls, shower curtains and tiles. They can also be seen hanging around outdoor septic systems. Their larvae live in pipes, drains and plumbing, feeding on the algae growing inside their surfaces.

Size: Adults 2–4 mm

Goodie or baddie? Adults are harmless but their presence in our bathrooms may indicate it is time to clean the drains!

Moth flies (*Psychoda* sp.) are tiny and can be found resting on the tiles and walls in our bathrooms. Source: L. Woodmore.

Other insects commonly found in the bathroom
>American cockroach, p. 80
>Ants, p. 174
>German cockroach, p. 79
>Silverfish, p. 256

Bedroom

Our bedrooms provide a wealth of food sources for hungry insects, from the woolly jumpers hanging in our cupboards, to our warm blood-filled bodies snuggled up in bed!

Bed bug

Cimex lectularius

Order Hemiptera

Ever heard of the saying 'Don't let the bed bugs bite?' Well, these bloodsucking critters really do inhabit bedrooms and drink our blood while we sleep. Fortunately, modern hygiene keeps them away, but they do crop up now and again in dodgy accommodation. During the day, they lurk in cracks and crevices around the room and in bedding. At night, when they sense our body heat and the carbon dioxide from our breathing, they crawl into our beds and start feeding. Bed bugs are tiny (about as big as a match head), oval-shaped and slightly flattened. They are usually reddish brown, but change to a much brighter red after a blood meal.

Size: Adults up to 5 mm

Goodie or baddie? Anything that sucks blood, especially while we are trying to sleep, is a baddie in my book! Their bites can cause itchy red welts on your skin and they may also leave little smears of bloody droppings on your sheets. Fortunately, bed bugs are not known to transmit diseases, unlike many other bloodsucking insects.

Bed bugs (*Cimex lectularius*) can be eliminated from your home through good hygiene practices, such as regular vacuuming. Source: Gilles San Martin, Attribution–ShareAlike 2.0 Generic.

Clothes moth
Tineola sp.
Order Lepidoptera

Clothes moths are small, secretive and rarely seen. Rather, it is the mess their larvae leave behind that usually reveals their presence. The larvae of clothes moths are very destructive, chewing their way through woollen fabrics, carpet, fur, feathers and clothes. They build tube-like cases from silk, droppings and strands of wool, which can still be found stuck to fabrics long after the moths have emerged and moved on.

Size: Adult wingspan 10 mm, larvae up to 10 mm long

Goodie or baddie? Clothes moths can cause lots of damage, especially to the woollen blankets and jumpers we have stored away during the warmer months.

Clothes moth larvae (*Tineola* sp.) construct tube-like shelters woven from silk and fibres to protect their delicate bodies.

Other insects commonly found in the bedroom
    American cockroach, p. 80
    German cockroach, p. 79
    Silverfish, p. 256

Around lights

Many insects are attracted to lights at night. You may notice them whizzing around the fluorescent lights on your ceilings, or see their undersides as they cling to your windows and screen doors, often dodging hungry geckos which move in for an easy meal. They also loiter around the bright lights near petrol stations or your favourite take-away restaurants. Many nocturnal insects use the moon as a reference point for navigation while flying. Artificial lights are much brighter, interfering with the insects' ability to detect the moon. This disorientates them and draws them to the stronger, artificial source of light.

Rhinoceros beetle

Xylotrupes ulysses

Order Coleoptera

Male rhinoceros beetles get their name from the large horns that adorn their head and thorax, a feature that the females lack. The adult beetles are generally found in the summer months, but can be seen all year round in far north Queensland. They have flying wings tucked underneath their shell-like wing covers and are often seen whizzing around lights by night and lumbering around on footpaths in the early morning. They also hang around poinciana trees, where they readily eat the bark on small branches. Males also gather here to battle each other for the chance to mate with a female, using their large horns to try to flip rival males off the branches. Both males and females produce a strange squeaking sound if you pick them up, made by rubbing their wings over grooves on their backs. This sound is designed to startle would-be predators.

Size: Adults 40–60 mm

Goodie or baddie? Apart from some sharp hooks on their feet rhinoceros beetles are harmless. However, their larvae may cause damage to the roots of certain plants (see curl grubs, p. 103), and the adults can ringbark ornamental trees.

Male rhinoceros beetle (*Xylotrupes ulysses*) showing the large horns on its head and thorax (which are absent in females).

Cane beetles
Many species
Order Coleoptera, Family Scarabaeidae

Also known as chafers, these brown or grey oval-shaped beetles often have a hairy or velvety appearance. Adults generally emerge in the warmer months after periods of rain and can be attracted to lights in large numbers. The adults and larvae feed on plants and can sometimes become a pest in our gardens.

Size: Adults up to 22 mm

Goodie or baddie? Their larvae (see curl grubs, p. 103) can damage the roots of plants and are serious pests in some commercial crops such as sugarcane.

This greyback cane beetle (*Dermolepida albohirtum*) is just one of the many species of cane beetles that are attracted to artificial lights.

Christmas beetles
Many species
Order Coleoptera, Family Scarabaeidae

Christmas beetles, as their name implies, are regular visitors to our lights around Christmas time; however, they can be found anytime during summer. They are usually brightly coloured, often metallic in appearance, and can be found feasting on the foliage of gum trees and wattles during the day. At night many are attracted to lights, but no longer in the huge numbers many of us recall seeing as children. Several causes may be responsible for their decrease, such as clearing of their habitat, the use of artificial lights in shops and streets that are designed to be less attractive to nocturnal insects, or perhaps a combination of both these factors.

Size: Adults up to 35 mm

Goodie or baddie? Christmas beetles can defoliate some trees when present in large numbers but, unless you are a commercial forester, they are a beautiful insect to have around your backyard.

As well as frequenting artificial lights at night, this Christmas beetle (*Anoplognathus porosus*) can be found feeding on the foliage of native trees during the day.

Mole cricket
Gryllotalpa sp.
Order Orthoptera

Mole crickets have distinct shovel-shaped fore legs, which they use to dig through soil. They live in deep underground burrows that open into a series of small galleries, some of which are used for egg-laying and rearing young. Within these burrows, they feed on the roots of plants and small soil-dwelling insects. Adult mole crickets can be heard singing at dusk – a rasping sound that is made by rubbing their wings together. They love moist conditions and are most vocal when it is raining or about to rain (and often after you water your lawn). The entrance to their burrows extends out like an old-fashioned gramophone horn, helping to amplify their songs. We most commonly encounter mole crickets when they are drawn to the brights lights around our patios and driveways at night.

Size: Length up to 40 mm

Goodie or baddie? The feeding behaviour of some species of mole crickets may cause minor damage to your gardens and lawn.

This mole cricket (*Gryllotalpa* sp.) is a juvenile and so lacks wings. Its shovel-shaped fore legs help it to burrow through soil.

Field crickets
Many species
Order Orthoptera

During the day, field crickets shelter under rocks and logs, within cracks in the soil and beneath vegetation. They come out at night (often aggregating around artificial lights) to feed and court each other. Male crickets sing by rubbing their wings together and can often be heard calling at dusk. If you get too close to them, they abruptly stop chirping and will only resume once you have moved away. These crickets are quite omnivorous, eating plant material, fruit and other small insects. They also scavenge on food scraps and dead animals.

Size: Length up to 30 mm

Goodie or baddie? Large populations of some species of field crickets can build up during certain times of the year, when their feeding can cause damage to your garden and lawn.

Field crickets, like this mottled field cricket (*Lepidogryllus* sp.), are a common sight around the bright lights of petrol stations where they gather to feed and court each other. Source: J. Dorey.

Katydids
Many species
Order Orthoptera, Family Tettigoniidae

Katydids are masters of camouflage. During the day, their leaf-like bodies enable them to remain virtually unnoticed in our gardens. However, at night they are often seen flying around lights or resting on our windows. Katydids sing at night, rasping their wings together in a similar fashion to their close relatives, the crickets. In fact, the name 'katydid' comes from the sound of the call of a species in North America. In some other species, these songs are pitched at a frequency that is beyond our range of hearing. Many katydids feed on the foliage of plants, although some species may prey on other small insects.

Size: Length up to 65 mm

Goodie or baddie? Some species can get into your veggie gardens, chewing large holes in your plants; however, many prefer the foliage of native trees.

During the day this katydid (*Caedicia* sp.) camouflages among the foliage of trees and shrubs, but they are often attracted to artificial lights at night.

Green lacewing
Mallada signatus
Order Neuroptera

These gauzy-winged insects are a beautiful sight on our windows at night, but be sure to look and not touch! When handled, green lacewings produce a foul-smelling liquid that is very difficult to remove from your hands. The females attach their eggs to the end of a long slender stalk of silk (see Chapter 5, p. 159), out of the reach of predators such as ants. Green lacewing larvae are voracious predators, eating many different types of insects. The adults feed on nectar and pollen and their green colouration allows them to hide on leaves during the day.

Size: Length 15 mm

Goodie or baddie? Lacewing larvae prey on many pest insects in your garden and are sold commercially in Australia as a biological control agent against many types of agricultural pests.

Green lacewings (*Mallada signatus*) can be found on the foliage of plants by day and are often attracted to artificial lights at night. Source: D. Papacek.

Other insects commonly found around lights
American cockroach, p. 80
Diving beetles, p. 93
Earwigs, p. 219
Giant water bugs, p. 95
Granny's cloak moth, p. 82
Indian meal moth, p. 76
Moths, p. 197
Termites, p. 265

In water

For insects, an underwater lifestyle has several advantages. The temperature in the water is often less variable than that of the surrounding environment, aquatic insects are safer from terrestrial predators such as birds, and there is a variety of food to choose from. Some insects, such as dragonflies and damselflies, use the water as a nursery, feeding and growing underwater until they are large enough to emerge on to land as winged adults. Others, such as diving beetles and backswimmers, spend their juvenile and adult lives in the water.

Some aquatic insects (see Table 2.7, p. 20) have very strict habitat requirements or are sensitive to water pollution and have therefore been excluded from this book. This section covers the more hardy species that frequently turn up in your local creeks, causeways, dams, ponds and, occasionally, your swimming pool.

Diving beetles
Many species

Order Coleoptera, Family Dytiscidae

Diving beetles trap little bubbles of air underneath their wing covers, allowing them to breathe underwater like a scuba diver. However, they frequently return to the surface to replenish their air supply before diving back down, hence their name. Their bodies are shiny, oval-shaped and streamlined for swimming. Their legs are flattened, oar-like and covered in long fringes of hairs to help them pull through the water. Adults and their larvae are voracious predators and feast on other aquatic animals in ponds, lakes and dams. The adult beetles have retained their ability to fly and often take to the air in search of new habitats. They are attracted to shiny surfaces which look like the water and are often drawn to artificial lights in large numbers.

Size: Adults up to 35 mm

Goodie or baddie? In the warmer months diving beetles can be attracted to lights in huge numbers, often making a nuisance of themselves.

The hind legs of this green diving beetle (*Onychohydrus scutellaris*) are fringed with long hairs to help propel its streamlined body through the water. Source: Gooderham and Tsyrlin (2002).

Whirligig beetles
Many species
Order Coleoptera, Family Gyrinidae

These small, oval-shaped beetles are shiny black, with long oar-like fore legs to assist in swimming. They can be found cruising around the water's surface, often in large groups. Their name comes from their habit of swimming around in crazy circles. Whirligig beetles have remarkable eyes that are split horizontally across the middle, with one half resting above the water's surface and the other half directed downwards, allowing them to patrol both domains simultaneously. Whirligig beetles are scavengers, feeding on insects that fall in the water and on the bodies of dead aquatic animals, such as small fish and crustaceans.

Size: Adults up to 18 mm

Goodie or baddie? These guys are the street-sweepers of our waterways, removing the bodies of dead aquatic animals through their foraging, which helps to recycle nutrients.

Whirligig beetles (Family Gyrinidae) are often seen swimming in circles on the water's surface, but they can quickly dive to escape danger. Source: Gooderham and Tsyrlin (2002).

Giant water bugs
Many species
Order Hemiptera, Family Belostomatidae

Giant water bugs are brown, leathery-looking insects, referred to as 'toe-biters' by some people. Their fore legs are armed with sharp hooks for hunting and their hind legs are fringed with hairs to aid in swimming. Both the adults and their similar-looking nymphs are ferocious predators and use a piercing needle-like mouth to feast on small fish, tadpoles and other aquatic animals. They can be found in ponds and lakes and can occasionally pop up in your swimming pool. Adult bugs are often attracted to artificial lights at night, as they fly in search of new aquatic habitats.

Size: Adults up to 75 mm

Goodie or baddie? If you have a fish pond, you may want to relocate any giant water bugs, as they will not hesitate to devour small fish and tadpoles. Don't pick them up with your hands as they are capable of inflicting a painful bite.

The giant water bug (*Lethocerus* (*Lethocerus*) *insulanus*) breathes air from the surface through a siphon at the end of its body.

Water striders
Many species
Order Hemiptera, Family Gerridae

Also referred to as pond skaters, water striders have the amazing ability to walk on the water's surface. Their mid and hind legs are very long and skinny and are covered with water-repelling hairs to prevent them from toppling into the water. Their fore legs are short and held close to their heads, giving them the illusion of having only four legs. Their bodies are small, brown and dull, and the nymphs look like smaller versions of the adults. They are often seen in large groups on the surface of ponds, lakes, slow-flowing streams and rivers, and occasionally in marine habitats. They use their piercing mouthparts to feast on aquatic animals, as well as spiders or flying insects that fall into the water.

Size: Bodies up to 8 mm, leg span of 50 mm

Goodie or baddie? Water striders are a lovely insect to observe in creeks and ponds, and they pose no threat to humans.

The water strider (*Tenagogerris euphrosyne*) uses its long legs to walk across the water's surface. Source: Gooderham and Tsyrlin (2002).

Water scorpions and needle bugs
Many species
Order Hemiptera, Family Nepidae

With their grasping fore legs and whip-like 'tails', you can understand how water scorpions get their name. They are not scorpions, they are true bugs, and the tail is not a stinger but a special snorkel used for siphoning air from the surface. It is located at the end of the abdomen, meaning these bugs can breathe air through their back end while their heads stay underwater, searching for prey. There are generally two different forms: needle bugs, which are long, lean and stick-like, and water scorpions, which have a broader, oval-shaped body. Both are brown and very well camouflaged, blending into submerged vegetation in the ponds and wetlands where they occur. They are clumsy swimmers; instead of swimming they sit and wait for fish or other aquatic animals to get within striking distance. Once seized, they use their piercing mouthparts to suck their prey dry.

Size: Adults up to 60 mm

Goodie or baddie? These insects can occasionally make their way into your swimming pool, which can be a nuisance. Don't pick them up as they are capable of biting when handled.

Needle bugs (*Ranatra dispar*) have long, stick-like bodies and strong sharp fore legs for grasping prey. Source: Gooderham and Tsyrlin (2002).

The leaf-like body of this water scorpion (*Laccotrephes tristis*) allows it to blend into its surroundings as it stalks its prey. Source: Gooderham and Tsyrlin (2002).

Backswimmers

Many species

Order Hemiptera, Family Notonectidae

Backswimmers and the similar-looking water boatman (Family Corixidae – not covered here) are found in the still waters of ponds, wetlands and sometimes even your swimming pool. Backswimmers are so named for their unusual swimming style. They swim on their backs, using their fringed hind legs to help them backstroke through the water. The bodies of the adults and nymphs are small and torpedo-shaped and their heads have large bulbous eyes. They are predators, feeding on a variety of small fish and aquatic animals. As they swim along on their backs, they keep an eye out for animals that have fallen into the water.

Size: Adults up to 10 mm

Goodie or baddie? Use a scoop to remove backswimmers from your pool – they can inflict a bite if handled roughly.

Backswimmers (*Anisops* sp.) can often be found in our swimming pools. Source: Gooderham and Tsyrlin (2002).

Dragonfly and damselfly larvae
Many species
Order Odonata (several Families)

Also referred to as nymphs or naiads, these larvae are the immature stages of winged dragonflies and damselflies. Damselfly larvae have long slender bodies and three large leaf-like gills at the end of their abdomens to help them breathe underwater. The larvae of dragonflies are shorter, stockier and breathe through tiny gills along the sides of their bodies. Both have large bulging eyes and tiny thread-like antennae, and are often green or brown to aid in camouflage. They are voracious predators and prey on aquatic invertebrates, tadpoles and even small fish. They can be found in a variety of habitats, from fast-flowing streams to stagnant ponds and dams. When the larvae reach a certain size, they leave the water and climb onto surrounding vegetation, where they split open their old skins to reveal the adult with its two pairs of wings. These shed skins (see Chapter 6, p. 218) can be found on trees and grasses growing near the water's edge.

Size: Larvae usually up to 30 mm, although some dragonfly larvae can reach 50 mm

Goodie or baddie? These quirky-looking larvae can help to keep mosquito wrigglers under control, but be sure to keep an eye out for them if you value the fish and tadpoles in your ponds.

Bluetail damselfly larvae (*Ischnura* sp.) inhabit the still waters of ponds, dams and lakes. Source: Gooderham and Tsyrlin (2002).

Darner dragonfly larvae (*Austroaeschna* sp.) can be found in streams in many parts of Australia. Source: Gooderham and Tsyrlin (2002).

Mosquito wrigglers

Many species

Order Diptera, Family Culicidae

'Wriggler' is the name given to the aquatic larvae of adult mosquitoes. They have long hairy bodies and large armoured heads, and can move quite quickly. Mosquito wrigglers loiter near the water's surface, sucking up oxygen from a siphon at the end of their abdomens. When disturbed, they quickly plunge down to the depths, seeking shelter. You may also see the comma-shaped pupae alongside the wrigglers, from which the adult mosquito emerges. Mosquito wrigglers prefer still, often stagnant waters such as swamps, wetlands and mangrove areas. They frequently turn up in artificial water bodies such as fishponds, birdbaths, dog water bowls, clogged gutters and drains and the bases of pot plants. Wrigglers use their bristle-like mouths to filter algae from the water; however, some species are predators and prey on the larvae of other mosquitoes.

Size: Larvae up to 10 mm

Goodie or baddie? While the wrigglers themselves are harmless, the bloodsucking adults of some mosquito species (see p. 135) cause itchy bites and are capable of transmitting viruses such as Ross River fever.

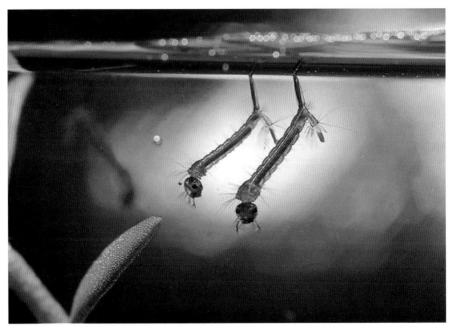

These mosquito wrigglers (Family Culicidae) are siphoning air from the water's surface. Source: Gooderham and Tsyrlin (2002).

Bloodworms

Chironomus sp.

Order Diptera

Bloodworms are not really worms, nor do they drink blood. Rather, they are the aquatic larvae of tiny mosquito-like flies known as non-biting midges, and the 'blood' refers to their bright red colouration. They have a red pigment in their bodies, which helps them to draw oxygen from the water. Bloodworms can be found in stagnant waters that contain pieces of vegetation, sediment and other detritus, upon which they feed. They often turn up in ponds, in containers of water around your garden (e.g. birdbaths and empty flower pots) and in the water-filled centres of plants such as bromeliads. Their bright red colour makes them easy to spot and they writhe around in the water by coiling and uncoiling their long worm-like bodies.

Size: Larvae up to 10 mm

Goodie or baddie? The larvae cause no harm and are an important food source for other aquatic animals. Unlike their close cousins the sand flies, adult non-biting midges (as the name suggests) do not bite humans.

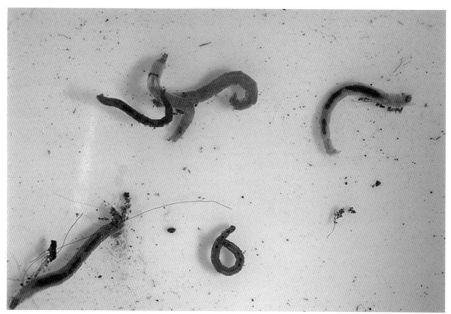

Bloodworms (*Chironomus* sp.) are the larvae of tiny mosquito-like flies. Source: Gooderham and Tsyrlin (2002).

Soil, leaf litter and compost

Soil, leaf litter and compost provide an excellent habitat for insects. They are cool and moist, offer a retreat from the heat and sun, and provide shelter from predators such as birds. Comprised of bark, leaf litter, fungi, roots and rotting vegetation, this habitat can be a smorgasbord for hungry insects.

Soldier flies
Many species
Order Diptera, Family Stratiomyidae

Soldier flies resemble wasps and have shiny black bodies and wings. In your garden, they can be found resting on the leaves of plants or near compost bins. The females use the pointy tips of their abdomens to deposit their eggs around the edges of the compost bin lid. Their larvae are brown, hairy and slightly tapered towards the head, which has no eyes. They feed on decaying organic material, such as vegetable scraps. The adult flies do not feed and are relatively short-lived.

Size: Adults and larvae up to 20 mm

Goodie or baddie? While they mimic wasps, soldier flies do not possess a sting. They do not buzz around and make a nuisance of themselves like house flies, and they play an important role in nutrient recycling. Perhaps the only negative is their tendency to show up in worm farms: while they do not directly harm the worms, they may breed large populations and compete for food and space.

A female soldier fly (*Hermetia illucens*) deposits her eggs around the edge of a compost bin.

Soldier fly larvae (*Hermetia illucens*) feed on decaying organic material.

Curl grubs
Many species
Order Coleoptera, Family Scarabaeidae

These large pale grubs are also known as white grubs, cockchafers and witchetty grubs and they are the larvae of certain types of scarab beetles, including rhinoceros, stag and cane beetles. We often dig them up in our gardens around the roots of trees and shrubs, in potted plants and compost bins. They have a very distinctive C-shaped body clothed in fine hairs, strong armoured heads and long legs for digging through the soil. They feed on decaying plant material, although some species eat the roots of living trees or turf. The really large ones you find in your garden most likely mature into rhinoceros beetles (see p. 86).

Size: Larvae up to 70 mm

Goodie or baddie? Some species can damage the roots of plants, especially potted plants and turf. Be careful if handling large larvae – they can give you a nip with their sharp mandibles.

This large curl grub is the larva of a rhinoceros beetle (*Xylotrupes ulysses*).

Hawk moth pupae
Many species
Order Lepidoptera, Family Sphingidae

If you dig around the soil of your garden, you may come across the pupae of hawk moths. The large hawk moth caterpillars feed on a variety of plants, including gardenias, impatiens and grape vines. They eventually drop down from the leaves and burrow into the soil where they form a pupa (pl. pupae) – the structure in which they transform into an adult moth. The pupae are shiny and brown, and can be as long as your thumb. They have a strange protrusion that looks like the handle of a jug. This structure houses the mouthparts of the adult moth, which eventually develop into a long coiled tube for slurping up nectar. The pupae may be covered in a flimsy cocoon of silk, or tucked neatly into a little compartment in the soil. If you pick them up, they wriggle and thrash from side to side.

Size: Pupae up to 70 mm

Goodie or baddie? The leaf-eating caterpillars of hawk moths can quickly strip the foliage from ornamental plants such as impatiens and gardenias, but the pupae do not feed and cause no damage to the roots of plants. The adult moths (see p. 204) are important pollinators of flowers.

Hawk moth pupae, such as those of the bee hawk moth (*Cephonodes kingii*), can be found buried in soil around the base of plants.

Wood roaches
Many species
Order Blattodea (several Families)

'Wood roaches' or 'burrowing cockroaches' are collective terms given to the many species of cockroaches that live outdoors in soil, under logs and in leaf litter. They are usually brown, with flattened oval-shaped bodies. Their spiky legs are used for digging, and adults may or may not have wings. They feed on wood, bark, leaf litter and other decomposing vegetation in the soil. Some species have special microscopic organisms in their guts to help them break down the tough cellulose in wood.

Size: Up to 30 mm

Goodie or baddie? Unlike their house-loving cousins, these cockroaches do not eat our food, spread germs or make a pest of themselves.

This wood cockroach (*Panesthia cribrata*) is common in compost heaps and underneath woodpiles.

Other insects commonly found in soil, leaf litter and compost
- American cockroaches, p. 80
- Antlions, p. 153
- Ants, p. 174
- Earwigs, p. 219
- Field crickets, p. 90
- Mole crickets, p. 89
- Termites, p. 265

On trees and shrubs

Many types of insects are found on the trees and shrubs that grow in our garden. Some are feeding – chewing holes in leaves, attacking fruit and ringbarking stems. Others are predators, hiding among leaves and branches as they stalk their prey.

Many herbivorous species feed on a particular type of plant throughout their life. For insects such as grasshoppers and aphids, both adults and nymphs can be seen feeding side by side on the one plant. In many butterfly and moth species the larvae are voracious leaf-eaters, while the winged adults feed on the sweet juices in flowers and fruit.

Aphids

Many species

Order Hemiptera, Family Aphididae

Aphids are tiny soft-bodied insects that may be green, yellow, brown or pale pink. They have long antennae that sweep back over their bodies and some adults have wings. The adults and nymphs feed side by side, targeting the growing tips of plants and sucking out the sap with their piercing mouthparts. They are found on many ornamental garden plants, weeds and fruit trees. Towards the end of their bodies is a pair of long slender (often black) tubes from which they excrete honeydew, a sugary waste product that is fed upon by certain species of ants.

Size: 1–2 mm

Goodie or baddie? Their sugary secretions can encourage the growth of sooty moulds on plants and may attract ants. Some aphids can transmit plant diseases through their feeding and cause wilting, stunted growth and plant deformities.

Rose aphids (*Macrosiphum* (*Macrosiphum*) *rosae*) target the shoots and buds of rose bushes. Source: D. Papacek.

Cicadas
Many species
Order Hemiptera, Family Cicadidae

Cicadas are stout-bodied insects with two pairs of wings (usually transparent) that are held roof-like over their bodies while at rest. They have a needle-like mouth, which they use to suck the sap out of plants (look for it tucked underneath their bodies between their legs at rest). While you can almost always hear cicadas singing in spring and summer, you may also see individuals resting on the trunks of eucalypts and other trees if you look carefully. The males use this perch to sing to females, a loud trilling, whining or buzzing sound that is made by contracting a stiff membrane on either side of the abdomen. Like birds, each type of cicada has a unique call so that only females of that particular species are lured by their serenading. You can find cicada shells (empty skins from the juvenile cicada) on the trunks of trees and on fences and posts around your home.

Size: Length up to 50 mm

Goodie or baddie? It depends on the numbers! While a few cicadas singing is very pleasant, some call in synchrony and hundreds of individuals, all clanging away, can be very irritating.

A female Dodd's bunyip cicada (*Tamasa doddi*) rests on the trunk of a tree, listening for the calls of courting males.

Cotton harlequin bug
Tectocoris diopthalmus
Order Hemiptera

Cotton harlequin bugs are beautiful metallic-coloured insects, often referred to as jewel bugs. Their shield-like bodies come in a range of colours and patterns. Females are larger than the males and are pale to dark orange, with splotches of metallic green on their backs. They can often be seen guarding their clusters of eggs on the stems of plants. Males are smaller and metallic blue, with shiny red and green patterns. The nymphs are smaller and rounder in shape, and resemble the males in colouration. Both adults and their young use their piercing mouthparts to suck the sap from plants, usually targeting the flower buds and soft shoots. Hibiscus, cotton plants and other closely related species make up their diet. They are most commonly encountered on the large beach hibiscus trees growing in parks and flanking footpaths near the ocean.

Size: Adult 20 mm

Goodie or baddie? These insects are a pest in cotton crops, where they attack the flower buds. However, in our gardens they do not cause significant damage and are a pretty sight. Just make sure you look and don't touch – these insects are a type of stink bug and let off an offensive smell if annoyed.

Adult female cotton harlequin bugs (*Tectocoris diophthalmus*) are commonly found on beach hibiscus trees.

Cotton harlequin bug nymphs (*Tectocoris diophthalmus*) are striking metallic blue, green and red. Source: James Niland, Attribution 2.0 Generic.

Assassin bugs
Many species
Order Hemiptera, Family Reduviidae

Assassin bugs are slow-moving predators that wander around shrubs, stalking their prey. Both adults and nymphs (p. 7) have slender legs, long thread-like antennae, large bulbous eyes and a short stout 'beak' that curves back under their head. After capturing their prey with their long fore legs, they inject digestive enzymes with their needle-like mouthparts, immobilising and liquefying their meal. It is then slurped back up through their 'beak', much like drinking a milkshake through a straw!

Size: Adult up to 25 mm

Goodie or baddie? Assassin bugs hunt many types of insects, including several pest species, making them a good addition to your garden. Do not handle them – they are capable of inflicting an extremely painful bite if annoyed.

Bee killer assassin bugs (*Pristhesancus plagipennis*) creep slowly around vegetation in search of prey.

Bush cockroaches
Many species
Order Blattodea, Family Ectobiidae (previously Blattellidae)

Unlike the pest cockroaches that raid our kitchens under the cover of darkness, beautifully coloured bush cockroaches can be found perched on the leaves and branches of trees and shrubs during the day. Their long spiky legs allow them to quickly scamper off if disturbed and their bright colours indicate they are distasteful to predators. The females often lay their eggs under bark, and the adults and similar-looking nymphs feed on plant material.

Size: Adults 10–20 mm

Goodie or baddie? Unlike their house-loving cousins, these native cockroaches do not spread germs, eat our food or make a pest of themselves.

Unlike most cockroaches, which shun the light, this beautiful ellipsidion cockroach nymph (*Ellipsidion* sp.) basks in broad daylight on a leaf.

Hedge grasshopper
Valanga irregularis
Order Orthoptera

The hedge grasshopper is a common visitor to our gardens and is one of the biggest grasshoppers in Australia. The winged adults are usually light brown or grey and may be coloured with blotches and spots. The nymphs lack wings and are usually pale green or brown, with patterns of black spots. Both have large eyes, long, usually black, antennae, and enlarged hind legs for jumping. The adults and nymphs have strong mandibles to chew through leaves and can be found feasting on your citrus plants, palm trees, hibiscus and other broad-leaved shrubs. When disturbed, the grasshoppers jump to safety. Adults also open their wings as they jump, allowing them to propel their bodies considerable distances.

Size: Adults up to 90 mm

Goodie or baddie? These grasshoppers can cause major damage to the foliage of your plants. If handled, the adults may lash out with their spiky hind legs and can bite with their sharp jaws if given the chance.

While this adult female hedge grasshopper (*Valanga irregularis*) is plain brown, some individuals may be patterned with dark spots or bands.

Longlegged flies
Many species
Order Diptera, Family Dolichopodidae

Longlegged flies, also referred to as green flies, are beautiful dainty insects. They have metallic green bodies, black bands on their backs and black spots on their wings. They are often found sitting on the broad leaves of plants and on the trunks of trees, but dart off if disturbed. They prefer shady moist gardens and their larvae develop in rotting vegetation or moist soil.

Size: Length 5–10 mm

Goodie or baddie? The adults hunt small insects, so may be useful in keeping pests (e.g. aphids and thrips) under control.

A longlegged fly (Family Dolichopodidae) rests on a broad leaf.

Crane flies
Many species
Order Diptera, Family Tipulidae

Crane flies look like giant mosquitoes but, thankfully, they do not suck blood. While there are over 700 species of crane flies within Australia, they can all be characterised by their elongated bodies and extremely long skinny legs. Many types have yellow or orange markings on their thorax and abdomen. You can find crane flies perched on broad leaves by day, or fluttering in the air at dusk. Adults can sometimes be seen dabbing the tip of their abdomen onto wet soil in the early morning, late afternoon or after watering the garden or a shower of rain. The maggot-like larvae live in water, moist soil and leaf litter, so the adults are most commonly found in shady, well-watered gardens.

Size: Length up to 75 mm, although usually around 20–30 mm

Goodie or baddie? The adults and their larvae pose no risk to humans or animals. The adults sometimes make their way indoors through open doors and windows.

Crane flies (*Nephrotoma australasiae*) are commonly found resting on broad leaves.

Leaf beetles
Many species
Order Coleoptera, Family Chrysomelidae

There are thousands of species of leaf beetles in Australia and they come in a variety of colours and sizes. Most are round or oval in shape (resembling large ladybird beetles), with a smooth, often shiny body. Some are solid green or brown, others have elaborate spots, stripes, or mottled markings and patterns. Their larvae (see p. 148) are caterpillar-like, with short thick bodies. They are often pale yellow, with black spots or stripes, and some are covered in dense hairs. Both the adults and larvae feed on leaves, often clustering together in large groups on a single leaf. They eat the foliage of a wide variety of plants, including native species such as eucalypts and wattles, ornamentals and the leaves of fruit trees.

Size: Adults 2–20 mm

Goodie or baddie? Some species of leaf beetles can defoliate garden plants completely, including fruit trees and vegetables.

A colourful eucalypt leaf beetle (*Paropsis maculata*).

Weevils
Many species
Order Coleoptera, Family Curculionidae

Many thousands of species of weevils exist in Australia and they all have something in common – a long 'snout' on their heads with a pair of stout elbowed antennae attached to it. While their size and colour vary enormously across species, most have a rounded body and a narrow head, giving them an oval or teardrop shape. Their bodies are extremely hard and are often highly patterned and textured. Many species can't fly and have a habit of suddenly dropping to the ground and feigning death very convincingly when disturbed. Their larvae are legless and white and are usually found in sheltered locations, such as within the tissue of plants or underground. The adults feed on plants, either devouring foliage or stripping the bark from stems and branches.

Size: Adults 2–65 mm, but usually less than 12 mm

Goodie or baddie? Some weevils can cause damage to plants through their feeding, although many native species restrict their diet to eucalypt and wattle trees.

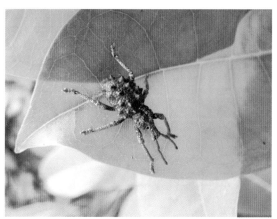

The body of this broad-nosed weevil (*Leptopius* sp.) is adorned with spiky-looking bumps.

Other insects commonly found on trees and shrubs

Ants, p. 174
Butterfly and moth larvae, p. 197
Christmas beetles, p. 88
Green lacewings, p. 92
Katydids, p. 91
Praying mantids, p. 250
Rhinoceros beetles, p. 86
Sawflies, p. 183
Spitfires, p. 127
Stick and leaf insects, p. 259

On citrus trees

Citrus trees such as lemons, oranges and limes are common in Australian backyards and for this reason they have earned their own special section in this chapter. The insects below frequently occur on our backyard citrus trees and some rely almost exclusively on these plants for food.

Bronze orange bug

Musgraveia sulciventris

Order Hemiptera

Bronze orange bugs are large, sap-sucking insects commonly seen in the warmer months, often in large numbers. The adults are dark brown to bronze. The young nymphs change from green to orange as they get older and have a characteristic black spot in the middle of their backs. They attack the shoots and young fruit of citrus trees, using their piercing mouthparts to feast on the sap. When disturbed, they spray a foul-smelling liquid that can burn your eyes and skin.

Size: Adults up to 25 mm

Goodie or baddie? These insects damage your citrus plants and, because of their noxious smell, are quite awkward to physically remove from fruit and foliage.

Bronze orange bugs (*Musgraveia sulciventris*) damage citrus trees through their feeding and produce a foul smell when disturbed.

Bronze orange bug nymphs (*Musgraveia sulciventris*) may be green, yellow or orange (depending on how old they are) and have a characteristic black spot in the centre of their bodies.

Spined citrus bug
Biprorulus bibax
Order Hemiptera

Spined citrus bugs earn their name for the prominent black-tipped spines on the 'shoulders' of their pale green bodies. The nymphs lack spines and have rounded green bodies, patterned with black, yellow and white markings. These bugs are a major pest in our gardens and have a preference for lemons, mandarins and oranges. They insert their needle-like mouthparts into the fruit and feast on the fluids, causing the fruit to drop prematurely.

Size: Adults up to 20 mm

Goodie or baddie? As well as damaging your fruit, these bugs omit a foul odour when handled.

Adult spined citrus bug (*Biprorulus bibax*) on a young orange.

Swallowtail caterpillars
Papilio sp.
Order Lepidoptera

Several species of swallowtail butterflies lay their eggs on citrus plants and they are good at finding even small isolated plants. Their caterpillars greedily munch on the leaves, but do not harm your fruit. The younger larvae are often shiny brown and white – mimicking little blobs of bird droppings! Some turn green as they get older, and are well-camouflaged. If disturbed, they can eject bright red inflatable 'horns' (see pp. 201, 206) from behind their head; these have a strong odour to deter predators. The pupae are green or brown and are attached to the stems of the plant, at an angle, with a little silk girdle.

Size: Adult wingspan 67–108 mm; larvae up to 70 mm

Goodie or baddie? These caterpillars do not damage your fruit. Although they can strip the leaves from your citrus trees, remember that they turn into a beautiful adult butterfly, which is a welcome sight in your garden.

Fuscous swallowtail caterpillars (*Papilio* (*Princeps*) *fuscus*) resemble bird droppings to deter predators.

This orchard swallowtail butterfly (*Papilio* (*Princeps*) *aegeus*) has just emerged from its chrysalis. Source: K. Hiller.

Other insects commonly found on citrus trees

Aphids, p. 106
Armoured scales, pp. 151–152
Assassin bugs, p. 109
Citrus leafminers, p. 149
Green lacewings, p. 92
Green vegetable bugs, p. 122
Hedge grasshoppers, p. 111
Mealybugs, pp. 152–153
Soft scales, p. 152

In the vegetable garden

Anyone who has tried to grow vegetables at home will know that many insects are adapted to eating the same plants as we do. Caterpillars and grasshoppers munch large holes through leaves and fruit. Sap-sucking bugs inject saliva into plants and suck up the resulting 'soup', causing shoots and buds to wither and die. However, think twice before you reach for the bug spray. Bees play a vital role in pollinating our plants and predatory insects help to keep pest species under control. Pesticides indiscriminately kill any species that are present – including beneficial ones which you would want to encourage in your garden. Instead, plant marigolds and chrysanthemums alongside your vegetables – they produce compounds that may repel some pest insects and their bright blooms draw in pollinators such as bees.

Cabbage white butterfly
Pieris rapae
Order Lepidoptera

Yellow and white butterflies are a pleasure to look at, but the larvae of cabbage white butterflies are unwelcome visitors in our gardens. Cabbage white butterflies are adept at locating suitable plants to lay their eggs on (mainly plants in the cabbage family) and their caterpillars feast on cabbage, nasturtiums, kale, brussel sprouts, broccoli and cauliflower. The caterpillars are green and very well camouflaged. They create large holes in the leaves of your plants and leave messy droppings as evidence of their feeding. The larvae eventually form a small green chrysalis (usually on the underside of leaves) and emerge as an adult butterfly after around two weeks.

Size: Adult wingspan 43 mm; larvae up to 30 mm

Goodie or baddie? Their camouflage makes it very challenging to find the cabbage white larvae and remove them before damage is done. While they prefer to munch on the leaves, their droppings also make a mess of your cauliflower and broccoli heads.

Cabbage white butterflies (*Pieris rapae*) are beautiful, but the destructive feeding of their larvae makes them unwelcome visitors to our vegetable gardens.

The larvae of cabbage white butterflies (*Pieris rapae*) are very well camouflaged.

Corn earworms and native budworms
Helicoverpa armigera, *Helicoverpa punctigera*
Order Lepidoptera

These two very similar species of moths are known as corn earworms, native budworms or heliothis moths (from their old Genus name). Both adult moths are light brown, with darker markings on their wings. As well as closely resembling one another in appearance, they have similar lifecycles and behaviour. They can migrate long distances in search of food plants for their caterpillars, and females can lay over 1000 eggs in less than two weeks. The larvae vary in colour from dark brown to light green, with longitudinal streaks (sometimes quite faint) and noticeable hairs on their bodies.

The caterpillars feed on a variety of vegetable crops, including sunflowers, soybeans and chickpeas, and attack the leaves, flower buds and seed pods of these and many other plant species.

Size: Adult wingspan 40 mm; larvae up to 50 mm

Goodie or baddie? The larvae of these moths can completely defoliate plants, making them one of the most destructive insects in Australian agriculture and backyard vegetable gardens.

Adult native budworm (*Helicoverpa punctigera*). Source: V. Dunis.

Corn earworms (*Helicoverpa armigera*) chew their way through leaves and developing fruit. Source: D. Papacek.

Green vegetable bug

Nezara viridula

Order Hemiptera

As the name suggests, green vegetable bugs are green and love to attack the vegetables in your garden. The adults have shield-shaped bodies and the nymphs, while similar in shape, are smaller with green, red, yellow and orange spots. Both the adults and nymphs use their long needle-like mouthparts (which are tucked under their bodies between their legs at rest) to pierce citrus fruits, beans and tomatoes, causing the shoots to wilt and fruit to drop prematurely.

Size: Adult 15 mm

Goodie or baddie? The adults emit a foul odour if disturbed and cause damage in vegetable gardens and crops, especially beans and snow peas.

The head and body of this green vegetable bug (*Nezara viridula*) is covered in small white eggs from a type of parasitic fly called a tachinid.

Brown bean bug

Riptortus serripes

Order Hemiptera

The brown bean bug is also known as the pod-sucking bug. Its body is long, slender, dark brown and narrow in the middle. There is a bright yellow stripe running along either side of its body and a pair of sharp spines on its 'shoulders'. The hind legs, which are much larger than the others, are lined with small spines and often splay out to the side. The nymphs (see p. 180) are also dark brown and look very much like ants. As the name suggests, this sap-sucking bug targets the filled pods of beans and peas in our vegetable gardens, sucking out the seed's juices with its needle-like mouth.

Size: Adult 18 mm

Goodie or baddie? These bugs cause significant damage and can reach large numbers during summer. They often shelter among dead leaves, making them difficult to spot, but they fly readily when disturbed.

Brown bean bugs (*Riptortus serripes*) are easily recognised by the spines on their 'shoulders' and the way they sit with their legs splayed out to the side.

Vegetable grasshopper
Atractomorpha similis
Order Orthoptera

The vegetable grasshopper is relatively small and green, with a long slender body and a pointed head, helping it resemble a blade of grass. It may also be brown, blending in among dead leaves and sticks. The hind legs are enlarged for jumping, which it does readily when disturbed. Adults have wings and the females are much larger than the males. The nymphs are similar in shape, but are smaller and lack wings. These grasshoppers have a preference for broad-leaved plants and chew holes through the soft leaves of your basil, mint, sage, lettuce and spinach.

Size: Adult up to 40 mm

Goodie or baddie? A minor pest in vegetable gardens. They are often difficult to spot and catch.

A mating pair of vegetable grasshoppers (*Atractomorpha similis*), showing the much larger female on the bottom. As this picture demonstrates, these grasshoppers come in both brown and green forms (regardless of their sex).

Other insects commonly found in the vegetable garden

>   Aphids, p. 106
>   Armoured scales, pp. 151–152
>   Assassin bugs, p. 109
>   Bees, p. 163
>   Field crickets, p. 90
>   Hedge grasshoppers, p. 111
>   Katydids, p. 91
>   Mealybugs, p. 152–153
>   Soft scales, p. 152

On native trees and shrubs

A vast number of Australian insects are adapted to feed, develop and reproduce on native plants such as eucalypts, wattles, tea trees and bottle brush. Some insect species are quite specialised, restricting their feeding to only a few native species and shunning exotic ornamentals. Others happily move between both. The following insects are commonly found on the trunks and foliage of native trees in parks, forests and our own backyards.

Eucalypt planthopper
Platybrachys sp.
Order Hemiptera

The eucalypt planthopper has a broad flat body and wings that are held roof-like over its back when not in use. The nymphs are wingless and smaller, with two long tails that poke out from the end of their abdomen. Both adults and nymphs are mottled brown and orange, helping them to blend into their surroundings. They use their needle-like mouths to feed on the sap of gum trees and can be spotted moving up and down the trunk in search of feeding sites. When disturbed, both adults and nymphs are capable of leaping into the air at lightning speed, a nifty trick for eluding predators.

Size: Adult 15 mm

Goodie or baddie? While some planthoppers attack fruits and vegetables, this species feeds on eucalypts and does not cause any noticeable damage to the tree.

An adult eucalypt planthopper (*Platybrachys* sp.) resting on the trunk of a gum tree. Both adults and nymphs can jump suddenly to escape danger. Source: Donald Hobern, Attribution 2.0 Generic.

Gum tree shield bugs
Theseus sp. and *Poecilometis* sp.
Order Hemiptera

Shield bugs are a type of stink bug and there are hundreds of different stink bug species in Australia. Gum tree shield bugs (of which there are several similar-looking species) are a common sight on eucalypts. The mottled black and greyish-brown adults can be found wandering around on the tree trunk, while the more striking black and white nymphs seek shelter under strips of bark. Both have bands of yellow and brown on their long thin antennae, shield-shaped bodies and slender legs for walking. Adults and nymphs feed on the sap of trees using their needle-like mouthparts.

Size: Adults 15 mm

Goodie or baddie? These bugs can let off a foul smell when disturbed.

This gum tree shield bug nymph (*Poecilometis* sp.) shelters under the bark of eucalypts. Source: James Niland, Attribution 2.0 Generic.

Spitfires
Many species
Order Hymenoptera, Family Pergidae

Spitfires are the larvae of sawflies, a type of primitive wasp. The caterpillar-like larvae have long bodies covered in short bristles and hairs, and three pairs of stumpy legs located near their heads. They are often seen feeding together in large clumps on native trees, such as eucalypts and paperbarks. As they feed, they extract powerful oils from the leaves and store them in their guts. When disturbed, the cluster of caterpillars simultaneously throw their heads back and spit out a yellow blob of this potent liquid, earning them the name 'spitfire'. At night, the clump disbands and spreads out to feed, maintaining communication by tapping their abdomens against the plant in a type of Morse code. As daylight breaks, they again join together in their defensive clump. They leave quite characteristic damage as a result of their feeding, stripping leaves from the top of the tree and gradually working their way downwards.

Size: Larvae up to 40 mm

Goodie or baddie? While other species of sawfly larvae attack fruit trees, spitfires usually target the foliage of native trees. Be careful – contact with spitfires can result in a painful burning sensation, leaving your skin red and irritated.

These sawfly larvae (*Perga* sp.) are known as 'spitfires' due to the droplets of burning fluid they spit from their mouths. Source: K. Ellingsen.

Other insects commonly found on native trees

- Bush cockroaches, p. 110
- Christmas beetles, p. 88
- Cicadas, p. 107
- Katydids, p. 91
- Leaf beetles, p. 114
- Praying mantids, p. 250
- Stick and leaf insects, p. 259
- Weevils, p. 115

In and around the lawn

A few select species of insects are found around the lawns in our backyards and on the grass bordering our footpaths. Some feed on the grass itself, whereas others are on the prowl for other insects to eat.

Lawn armyworm

Spodoptera sp.

Order Lepidoptera

Lawn armyworms are the larvae of a greyish-brown moth. The caterpillars are sausage-like, with long dark stripes running the length of their bodies. They usually start off green and gradually darken to brown as they get older, when they develop distinct black triangular markings on their bodies. If disturbed, they roll up into a tight coil. The adult moths lay large clusters of eggs covered in fine light brown scales, giving them a furry appearance. Over 1500 eggs are fastened to walls, brickwork and the underside of trees growing near the grass. The larvae feed under the cover of darkness, stripping the soft edges from the grass blade and often leaving the tough midrib behind. During the day, they can be found sheltered near the base of the grass tussocks. If you leave lawn clippings behind when you mow your lawn, you can often find the larvae sheltered underneath the next day.

Size: Adult wingspan up to 40 mm; larvae up to 45 mm

Goodie or baddie? The secretive nocturnal feeding of lawn armyworms means they often go unnoticed until they have caused major damage to your turf.

Lawn armyworms (*Spodoptera* sp.) often shelter under piles of lawn clippings.

Orchid dupe
Lissopimpla excelsa
Order Hymenoptera

The orchid dupe is a beautiful slender wasp, with a bright reddish-orange body and legs and glossy black wings. The name 'orchid dupe' comes from an unfortunate mistake often made by the male wasps. Certain types of orchids produce flowers that are similar in shape and smell to female wasps. Love-struck males mate frantically with these false females, helping to pollinate the plant in the process. The females are a common sight in suburban gardens, flying low over our lawns during the day. They use a long 'needle' at the end of their abdomen to probe among grass tussocks in search of grubs such as the lawn armyworm (see p. 128). Once she has located a caterpillar, the female injects her eggs inside its body. From the eggs hatch maggots, which feast on the internal tissue of the caterpillar, eventually killing it. Insects with this type of lifecycle are called parasitoids.

Size: Adults up to 35 mm

Goodie or baddie? The females help to control pesky lawn grubs and the males are important pollinators of certain orchids.

An orchid dupe wasp (*Lissopimpla excelsa*) prowls the lawn in search of grubs inside which she lays her eggs. Source: J. Dorey.

Blue-ant

Diamma bicolor

Order Hymenoptera

A blue-ant is not an ant at all, but a female wasp that emerges from her pupa without wings. Their bodies are a beautiful metallic blue-green and the legs are bright red or orange. Unlike some wasps, blue-ants are solitary insects and do not form nests or hives. The females can be found wandering around on the ground and can move quite quickly. They are searching for mole crickets (see p. 89), which live in small chambers underground. Once she has located her prey, the female immobilises it with a sting and injects her egg into its body. Her maggot-like larvae feed and develop inside the body of the cricket. Male wasps are smaller and have wings, and can be found in flowers where they feed on nectar.

Size: Up to 25 mm

Goodie or baddie? As pretty as they are, stay well clear of blue-ants. The females have a very painful sting and can cause allergic reactions in some people.

Blue-ants (*Diamma bicolor*) are not actually ants, but a type of wingless wasp. Source: J. Dorey.

Greenhead ant
Rhytidoponera metallica
Order Hymenoptera

Also commonly referred to as green ants, greenhead ants have a reputation for ruining picnics and outdoor sports. These ants are actually quite beautiful in appearance, with metallic blue, green and purple bodies. However, a sharp stinger at the end of their abdomens makes them an unwanted resident in our backyards and local parks. They build their nests deep within the soil. The entrance to their nest may be visible in our grassy lawns, or may remain hidden underneath rocks and logs. The ants quickly appear to scavenge fallen crumbs at picnics and have a nasty habit of stinging you on the backs of your legs as you sit on the grass.

Size: Up to 6 mm

Goodie or baddie? Greenhead ants are one of the most commonly encountered stinging ants in Australia. Their stings can be extremely painful, with a sharp throbbing sensation that can last for several hours and leave behind an itchy red welt.

This greenhead ant (*Rhytidoponera metallica*) is very beautiful with its metallic purple and green body, but its sting really packs a punch. Source: J. Dorey.

Other insects commonly found in and around the lawn

Curl grubs, p. 103
Field crickets, p. 90
Mole crickets, p. 89
Pasture funnel ants, pp. 153–154

In and around flowers

Some flowers contain swollen vessels of sweet nectar, others produce large quantities of pollen. Nectar provides an easily digested, carbohydrate-rich fuel for many flying insects. Pollen is also highly sought after, as it is loaded with important nutrients including protein, vitamins and minerals. Many flowers rely on foraging insects for pollination and advertise their spoils with bright coloured petals, strong and sweet odours, and ultra-violet reflecting spots that illuminate a path to the stores of nectar and pollen.

Meat ants

Iridomyrmex sp.

Order Hymenoptera

Meat ants are not merely carnivores, as their name suggests. They are omnivorous and have a wide diet that includes other insects, larger animals such as frogs and lizards, seeds and plant material, honeydew from sap-sucking insects and nectar from flowers. These large, reddish-brown ants can often be found foraging on the flowers of native trees such as eucalypts. The nest of a meat ant consists of a huge underground gallery, filled with networks of tunnels and chambers.

A large mound of soil marks the location of the nest, which is usually situated in an area of dirt that has been cleared of vegetation. This dome of soil is covered in small pebbles, gravel and sometimes dead leaves, and contains several entrance holes.

Size: 8 mm

Goodie or baddie? These ants do not possess a sting, but can give you a nip with their strong jaws.

Meat ants (*Iridomyrmex* sp.) have a very diverse diet which includes nectar from blossoms.

Hover flies
Many species
Order Diptera, Family Syrphidae

Hover flies do just that – they hover in the air. Although they appear to hang motionless, they can dart away quickly if disturbed. Their slender bodies often boast black and yellow patterns, closely resembling bees and wasps. However, if you look closely you will see they only have one pair of wings – a feature unique to flies. The adults dart between flowers on sunny days, feasting on nectar. Their maggot-like larvae can be found on plants, hunting soft-bodied insects such as aphids and devouring them.

Size: 10 mm

Goodie or baddie? Despite their resemblance to wasps, hover flies do not sting. The adults are important pollinators of flowers and the larvae prey on pest insects in your garden.

A black-headed hover fly (*Melangyna* sp.) drinks nectar from a flower.

Flower beetles
Many species
Order Coleoptera, Family Scarabaeidae

Also known as flower scarabs or flower chafers, these beetles come in a variety of sizes and a dazzling array of striking colours and patterns. These powerful fliers seek out flowers on warm sunny days and often buzz loudly as they zoom past. They are found mainly on the flowers of native trees and shrubs, such as eucalypts and tea trees. Many species feed alongside one another on a cluster of flowers, with their heads thrust into the blooms as they feast on nectar and pollen. Some species of flower beetles breed up huge populations in the summer, often landing on our washing as they swarm in search of flowers.

Size: Up to 32 mm

Goodie or baddie? The beautiful colours and pollinating skills of these beetles make them welcome visitors to our gardens. Some plague-forming species (such as the species illustrated) can be a bit bothersome as they land on our washing or blunder into our faces and hair.

Green scarab beetles (*Diphucephala* sp.) feed on the foliage and flowers of a variety of plants and can mass in large populations at certain times of the year.

Other insects commonly found in and around flowers
Bees, p. 163
Butterflies and moths, p. 197

On the bodies of animals – the bloodsuckers

Many insects live and feed on the bodies of other animals. Lice spend their entire lives on the body of their host and have special adaptations, such as flattened bodies and strong claws on their legs, to help them cling to feathers and fur. For other insects, blood feeding is only undertaken by the adults, with the young developing in an entirely different habitat. Mosquitoes are one such example, with their aquatic, plant-feeding larvae eventually growing into flying bloodsucking adults. Blood is an easily obtained source of protein, which is an important nutrient in the development of eggs in females.

Mosquito

Many species

Order Diptera, Family Culicidae

Mosquitoes are one of the most commonly encountered bloodsucking insects in our homes and gardens. There are around 300 species in Australia and most are black or brown, with long slender legs. Mosquitoes are actually flies and, like all flies, have only one pair of wings. Their bodies are slender and they have mouthparts modified into a piercing 'syringe' for siphoning blood from animals. Only female mosquitoes drink blood, as they use the proteins and minerals within it to help develop and mature their eggs. The males feed on nectar and plant secretions and are totally harmless to humans. The larvae of mosquitoes (see p. 100) are also harmless, feeding on decaying plant material in aquatic habitats.

Size: Adults up to 9 mm

Goodie or baddie? The bloodsucking females of some species cause itchy bites and are capable of transmitting viruses such as dengue fever and Ross River fever. Town councils often 'fog' insecticides into gardens, wetlands and other mosquito habitats to control them.

Only female mosquitoes, such as this grey striped mosquito (*Ochlerotatus vittiger*), drink blood. Source: J. Dorey.

Biting midges
Many species
Order Diptera, Family Ceratopogonidae

Biting midges, also known as sand flies, are minute black or brown flies. Their tiny size means they often go unnoticed by humans – that is, until they start to bite! Biting midges have a very painful bite, which can leave behind large, itchy red welts that may take several days to disappear. Like mosquitoes, it is the females which suck the blood of warm-blooded animals, while the males are harmless to humans. The larvae develop in mud, wet sand and brackish waters, and so we often encounter the adults around coastal areas.

Size: Adults up to 3 mm

Goodie or baddie? During the warmer months, these insects can breed in large numbers and humans are frequently bitten. As the name 'sand fly' suggests, these bloodsucking insects frequent coastal areas and can negatively affect our time spent on or near the beach.

Despite their tiny size, the bites of biting midges (*Culicoides* sp.) can be very painful. Source: Science Image.

Horse flies
Many species
Order Diptera, Family Tabanidae

Also known as March flies, horse flies are common visitors to our gardens in the warmer months. Their bodies are short and stout and their enormous shimmery yellow-green eyes take up almost their entire head. The females feed on the blood of animals, including livestock and humans, while the males prefer to drink nectar from flowers. They are very fast fliers and often gently settle on our skin without us noticing, denying us the chance to shoo them away before they begin to bite. The feeding method of the females makes their bites especially painful. Instead of injecting a fine needle and sucking up the blood like a mosquito, horse flies use a thick stabbing mouth to scratch at your skin then lap up the blood as it pools in the wound.

Size: Adults 6–20 mm

Goodie or baddie? These flies are perhaps one of the most annoying uninvited guests to have at a picnic or BBQ. They buzz around your body and legs and insect repellent does not seem to deter them. Many of us swat a horse fly after receiving a painful bite, only to watch them shake themselves off before mounting another attack.

This horse fly (Family Tabanidae) is about to plunge its stabbing mouthparts into an unfortunate victim's thumb.

Other insects commonly found on the bodies of animals
Bed bugs, p. 84
Fleas, p. 221
Lice, p. 246

In large groups – masses and migrations

At certain times of the year, usually in the warmer months, the size of some insect populations can explode. A combination of favourable weather conditions and an abundance of food can send the rates of breeding and egg-laying skyrocketing and, before you know it, tens of thousands of insects descend upon our homes and gardens.

Other insects, such as some types of butterflies, migrate in large masses – flying long distances in a purposeful manner, ignoring food and potential spouses as they move towards their destination. The end point may be a breeding site where there is an abundance of plants on which to lay eggs, or it could be a place for over-wintering – a warm, sheltered location where adults cluster together to wait out the coldest months of the year.

These large 'plagues' are often witnessed by the public and sometimes attract the attention of the media. Some species are admired as they gather together in spectacular masses, others may cause damage to crops and gardens. While these fluttering masses of insects may invade our personal space for a short time (e.g. flower beetles clustered on washing, butterflies splattering on car windscreens, plague locusts blundering into our face and hair), it is a mild annoyance compared to the excitement of witnessing such a grand and extraordinary phenomenon.

Processionary caterpillars

Ochrogaster lunifer

Order Lepidoptera

Processionary caterpillars are the hairy larvae of the bag-shelter moth, a fluffy pale brown moth with small white markings on its wings and a long furry abdomen. Both the adults and larvae are clothed in dense tufts of itchy hairs which, upon contact, can cause painful rashes and nasty allergic reactions. The caterpillars have voracious appetites and can completely defoliate the wattle trees they feed upon. This is when individuals form the long processions their name refers to, marching off head to tail in search of more food. This behaviour is most common in late summer and early autumn, when the hungry caterpillars are at their largest. When they are not feeding, they rest in a silken bag swathed in itchy hairs, either at the base of the tree or high up in the foliage.

Size: Adult wingspan up to 65 mm; larvae up to 50 mm

Goodie or baddie? As well as causing painful skin rashes in humans, accidental consumption of these itchy hairs by grazing horses has reportedly triggered the loss of unborn foals in mares.

HABITAT – WHERE INSECTS LIVE AND OCCUR 139

A long trail of processionary caterpillars (*Ochrogaster lunifer*) marches off in search of new host plants. Source: Entomology.

Caper white butterfly

Belenois java

Order Lepidoptera

Caper white butterflies are handsome white insects with yellow and black markings. They can form large migrations in spring and summer, which have been dubbed by some as 'butterfly snow'. At its peak, as many as 650 individuals have been witnessed passing through a 50 m line of sight within the space of an hour. The butterflies move at a steady speed, around 2–3 m off the ground. Unfortunately, at this height, many individuals become splattered on our car windscreens.

The exact nature of these migrations is quite baffling and the direction of flight often depends on your locality. Queensland populations tend to fly further north during the spring, whereas those in New South Wales and Victoria fly south.

Caper whites breed in vast numbers on caper bush, a shrub that grows commonly, further inland from the coast of Australia. The larvae reach huge populations, often completely stripping the foliage from the plant. The migrations of the adults are almost certainly linked with the butterfly's search for suitable plants for egg-laying, as well as warm mild temperatures for optimal survival.

Size: Adult wingspan 55 mm

Goodie or baddie? These migrations are a beautiful sight, but a trip through the car wash may be needed.

A thirsty caper white butterfly (*Belenois java*) pauses for a drink of nectar.

Blue tiger butterfly
Tirumala hamata
Order Lepidoptera

Blue tiger butterflies are large dark brown or black butterflies, with streaks of light blue markings dotted across their wings. They can be found breeding in coastal and monsoonal rainforests, where their caterpillars feed on certain species of vines. From time to time they form huge migrations during spring, usually making their way southward. In autumn, the butterflies fly north in search of over-wintering sites, where they wait out the cooler months huddled together on trees and vines in dry gullies and creek banks.

Size: Adult wingspan 70 mm

Goodie or baddie? A beautiful butterfly to watch flutter by.

Blue tiger butterflies (*Tirumala hamata*) jostling for a feed of nectar.

Lantana treehopper

Aconophora compressa

Order Hemiptera

Lantana treehoppers are tiny sap-sucking bugs. They have a small horn protruding above their head, so that their bodies resemble the thorns on a tree. The adults are brown and the nymphs, while similar in shape, are paler and smaller and have black stripes running down their bodies. These insects were introduced to Australia from Mexico in 1995 to help in the biological control of lantana. However, since their introduction they have moved onto other trees and shrubs, including fiddlewood, geisha girls and jacarandas. The adults and nymphs plunge their piercing mouthparts into the plant and suck out the sap, causing die-back and defoliation. During the colder months, large masses of treehoppers cluster together on the branches of fiddlewoods. However, they cannot tolerate hot temperatures and their numbers drop dramatically in summer.

Size: Adults 8 mm

Goodie or baddie? These insects produce honeydew, a sugary waste product that can promote the growth of sooty moulds and will stain patios and cars parked underneath infested trees. However, the detrimental side effects of their sap-sucking diet does play some role in controlling lantana – a noxious weed that invades many forests and creek lines.

Lantana treehoppers (*Aconophora compressa*) resting on the branch of a lantana shrub.

Australian plague locust
Chortoicetes terminifera
Order Orthoptera

Locusts are simply large species of grasshoppers, well known for their swarming. They are green or brown, with dark brown splotches on their wings and large spiky hind legs. The nymphs are similar in appearance, but are smaller and lack wings.

After periods of rain in typically dry country (e.g. the interior of Australia), favourable conditions cause populations to explode and these normally solitary insects start aggregating in large ravenous swarms. These 'plagues' of locusts spread through the countryside in search of food. A typical locust swarm can contain over a billion individuals, weigh many tonnes and migrate distances of up to 1000 km.

Size: Adults up to 32 mm

Goodie or baddie? The Australian plague locust is the most economically significant grasshopper in Australia as it causes millions of dollars worth of damage to crops and pastures. When a plague descends upon a town it can disrupt sporting events, devastate gardens and make driving and other outdoor pursuits almost impossible.

Australian plague locust (*Chortoicetes terminifera*). Source: L. Woodmore.

Other insects that may form masses and migrations
 Flower beetles, p. 134

5
Clever clues – the strange structures and evidence that insects leave behind

Insects are not always easy to see, but they may leave behind characteristic evidence of their feeding, growth and reproduction that can be used in their identification. For example, herbivorous insects may blemish leaves with holes and scars as they feed, and some wasps construct elaborate mud nests for their young. Other insects hide beneath armoured plates, among frothy bubbles of fluid or inside fleshy bumps on plants – their resulting appearance is so bizarre that we don't actually realise they are an insect. This chapter helps you to identify cryptic insects by examining their strange structures and the evidence they may leave behind.

Using this chapter

From the categories listed in Fig. 5.1, choose the type of evidence your mystery insect has left behind or select the strange structure you suspect may harbour an insect. This refers you to an information page where these 'clues' have been divided into more specific subcategories. From here, you will find information on the insect or insects that are responsible for each type of clue.

For example, if you choose 'markings on leaves or bark' in Fig. 5.1 and refer to p. 146, you will find photographs of the different types of markings that are commonly found on leaves and bark (e.g. semi-circular holes around the leaf margin), along with the insect or insects responsible for causing such damage. You can read in Chapter 6 about the Order that the insect belongs to, to learn more about its general lifecycle, habitat and survival tactics.

Tips for getting started

Before you get started, here are some tips to help you navigate through this chapter.

Tip 1. Choosing the type of clue

It can be difficult to decide exactly what type of clue your insect has left behind. For example, you may decide you have found a nest (p. 153), but you could have it confused with a cocoon or egg case (p. 155). At the end of each section, you are referred to other places where your type of clue may be listed, in case you have assigned the clue into the wrong group.

Tip 2. Making the cut

There is a myriad of evidence and strange structures insects can leave for us to find, and any one of hundreds of different insect species may be responsible for any given clue. For example, some bumps and swellings on plants are known as galls and they are caused by several insects, across several different Orders. Many galls are formed by other organisms, such as bacteria and fungi.

The focus of this chapter is on recognising the main Orders of insects responsible for a particular clue, rather than tracing a clue back to the exact species of insect that has caused it. In some instances, specific insects have been listed for a certain clue, but only if the clue in question is caused by that insect alone.

Tip 3. Can't find it here?
Did you not find your particular insect in this chapter? Keep an eye on the structure or evidence that has been left to see if the insect responsible for it returns. For example, if you find some sort of cocoon or egg case, place it in a jar and see if anything emerges from it. If you can get to see the insect culprit, you can use the identification key (Chapter 3) to identify your insect based on its appearance.

If the same type of evidence keeps popping up in a particular location around your home or garden, the habitat chapter (Chapter 4) may be useful in identifying the insect responsible for it. For example, if you are finding holes and bite marks on your lemon tree, you can look up 'On citrus trees' (p. 116) and refer to an information page where the insects most commonly encountered on citrus are listed in detail.

Choose the clue
Select the clue that your insect has left behind and refer to the corresponding section.

Markings on leaves and bark
Many insects feed on plants, but are well camouflaged or have secretive behaviours to help avoid detection by enemies. Others aggregate and feed together in large visible groups, but move on before we can catch them in the act. A closer examination of the types of damage caused by the feeding behaviour of herbivorous insects can be helpful in their identification.

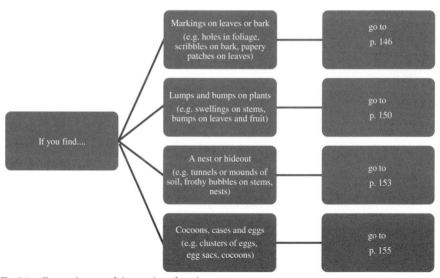

Fig. 5.1: Choose the type of clue you have found.

Holes and bites on leaves

To dine on leaves, insects must be equipped with strong sharp mandibles to cut through their food and a sturdy digestive system to help break down tough plant fibres and sometimes the toxic compounds found in foliage. For this reason, the defoliation of a particular plant type is usually caused by one of only a handful of insect species. Grasshoppers and stick insects (both adults and nymphs) and the caterpillars of both moth and butterfly species devour leaves, but their knack for camouflage means that even a thorough search of a plant will not always reveal their presence. The adults and larvae of beetles and the larvae of sawflies also defoliate plants but they often aggregate in large groups, making them easier to spot as they feed (unless they are high in the canopy).

Insects that may leave holes and bites on leaves

> Beetles and their larvae, p. 186
> Butterfly and moth larvae, p. 197
> Grasshoppers and crickets, p. 233
> Sawfly larvae, p. 183
> Stick and leaf insects, p. 259

Semi-circular holes around leaf margin

If the leaves of your rose bushes look as though someone has attacked them with a hole-punch, it is most likely the work of a crafty native bee. Leafcutting bees, as their name suggests, use their sharp mouthparts to cut neat circular or oval-shaped holes through leaves. These solitary bees then fly these disks back to their homes, usually located in small cavities such as hollowed-out plant stems or the gaps around your

Trees and shrubs (including native species, ornamentals and fruit and vegetable plants) may have large holes in the centre of their leaves and bite marks around the leaf margins.

Sometimes the insects responsible for large bites and holes in leaves are very elusive. This moth caterpillar (marked with an arrow) is almost invisible among the foliage upon which it feeds.

window sills. Your butchered rose leaves are moulded into plugs to form individual cells, which are stocked with pollen for the developing young and capped off with a neat circular plug of leaf. The damage caused by leafcutting bees is easy to distinguish from that caused by other leaf-eating insects as the cut margins are very smooth and neat, unlike the ragged edges left by caterpillars and grasshoppers. Although it can look unsightly, the leaf damage is a small price to pay for the pollination services of these bees.

Lacy or skeletonised leaves

Some insects are fussy eaters, avoiding the tough leaf veins and feasting only on the soft tissue in between. This causes a very distinct lacy pattern in leaves. The main culprits causing this kind of damage are beetles and their larvae, especially leaf beetles belonging to the family Chrysomelidae. Certain types of moth and sawfly larvae also skeletonise leaves.

Insects that may skeletonise leaves

> Beetle larvae, p. 186
> Butterfly and moth larvae, p. 197
> Sawfly larvae, p. 183

The leaves of this rose bush have neat semi-circular holes with smooth edges.

Figleaf beetle larvae (*Poneridia australis*) aggregate to feed on the soft tissue between the veins of this fig leaf, giving it a lacy or skeletonised appearance. The larvae will eventually turn into small yellowish-brown beetles.

Leafcutting bees (*Megachile* (*Eutricharaea*) sp.) remove circular disks from leaves to be used in the construction of their nests. Source: L. Woodmore.

Papery patches on leaves

Thin, single-layered papery patches commonly found on the leaves of eucalypts are the work of the leafblister sawfly, which is not actually a fly but a type of wasp (see Chapter 6, p. 183). The female inserts her egg

CLEVER CLUES – THE STRANGE STRUCTURES AND EVIDENCE THAT INSECTS LEAVE BEHIND

Papery patches or blisters can be found on the surface of eucalypt and wattle leaves.

Silvery scribbles on the leaf of an orange tree are the work of a citrus leafminer (*Phyllocnistis citrella*). This tiny caterpillar will eventually turn into a small white moth. Source: D. Papacek.

Adult leafblister sawflies (*Phylacteophaga* sp.) lay their eggs into leaves and it is their caterpillar-like larvae that create the papery patches, under which they feed and grow. Source: L. Woodmore.

into the leaf and from it hatches a tiny caterpillar-like sawfly larva. It burrows beneath the upper surface of the leaf, causing a small blotch or blister on the leaf surface, while it feeds on the green tissue beneath. This blister increases in size as the larva grows and it soon turns brown, making the leaf look as though it is dead. The caterpillar turns into a pupa within the blister, from which the adult eventually emerges. Adult sawflies feed on nectar, not leaves, and so do not cause any further damage to the plant.

Scribbles on leaves

Tiny insect larvae are typically responsible for the strange silvery squiggles sometimes found on leaves. Certain species of moths, sawflies, wasps, beetles and flies deposit their eggs on or under the surface of leaves. Their larvae burrow between the upper and lower surfaces of the leaf, forming a narrow tunnel in which they feed and shelter. The larvae usually have a flattened body to help them live in such a tiny space. The larva usually turns into a pupa within the tunnel, before emerging as a winged adult.

Insects that may leave scribbles on leaves

Beetle larvae, p. 186
Butterfly and moth larvae, p. 197
Fly larvae, p. 225
Sawfly larvae, p. 183
Wasp larvae, p. 169

Scribbles on the trunks of gum trees

The curious patterns adorning the smooth bark of scribbly gums are the work of a crafty moth caterpillar. The larva burrows under the surface of the bark, forming a winding zigzag pattern as it devours the plant tissue. You can actually track the growth of the larva by following the patterns – you can see where it has moulted its skin to a larger size, as the tunnel widens at each such point. When the caterpillar is roughly half-grown, it does a

U-turn and doubles back parallel to the original tunnel. It drops to the ground and forms a pupa in the leaf litter before eventually emerging as a small grey moth.

Can't find it here? The markings you have found may be listed as a 'Lump or bump'. Refer to the next section.

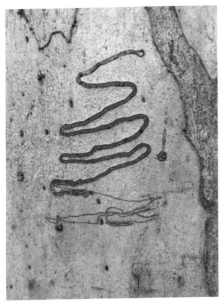

The trunks of certain gum trees are covered in brown scribbles and squiggly patterns, earning them the name 'scribbly gums'.

The larvae of scribbly gum moths (*Ogmograptis* sp.) leave scribbles in the bark as they burrow through the tissue of the tree. Source: N. Barnett, CSIRO.

Lumps and bumps on plants

Often we find strange lumps and bumps on our plants. These are often the handiwork of insects. Sometimes these formations are the insects themselves, or they may be a structure made by the insect for food, shelter or protection from enemies. Be careful, though, as other organisms (e.g. bacteria and fungi) can also make such structures.

Swellings, spikes and spheres on stems and leaves

Galls are strange bumps and swellings on the leaves and stems of plants. Bacteria, fungi, nematodes (a type of worm), mites and many types of insects can produce galls. Galls are the plant equivalent of a tumour. The feeding of an insect can irritate the plant, causing it to rapidly lay down new cells in an attempt to repair the damage. This eventually develops into an abnormal mass of tissue, known as a gall, and the distorted pattern of growth may be used to the advantage of the insects inside. Gall-forming insects shelter inside these little structures and consume the modified tissue within. Certain types of flies, wasps and sap-sucking bugs such as psyllids, aphids and mealybugs can all form galls.

A gall on the stem of a native tree.

Insects that may leave swellings, spikes and spheres on stems and leaves

Fly larvae, p. 225
Psyllids, aphids and mealybugs, p. 278
Sawfly larvae, p. 183
Wasp larvae, p. 169

Tiny shells, delicate plates or feathery structures on leaves

Many types of eucalypts and wattle trees have the surface of their leaves adorned with tiny pale scales or plates. Up close, these structures can resemble delicate seashells or little feathers, firmly stuck to the leaf. These structures are known as lerps and are formed by a true bug known as a lerp insect. There are many different species, each with its own characteristic lerp design. Lerp insects use their needle-like mouth to feed on plant sap. The nymphs construct their home using wax and honeydew, a waste product excreted from their bodies. They shelter under this protective covering until they reach adulthood.

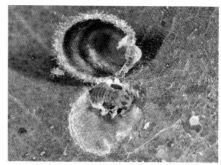

Prising off the shell-like plate on a leaf reveals the tiny lerp insect (*Hyalinaspis* sp.) beneath. Source: P. Chew.

Hard plates or bumps on leaves and fruit

Minute sap-sucking bugs known as hard or armoured scale insects produce hard plates and shiny bumps on plants. These insects initially start life as small, highly mobile nymphs called crawlers that wander around a plant until they find a suitable place to feed. Once the nymph finds a good location, it lays down layers of wax over its body, forming a protective armour against the elements. This covering can be circular, oval or teardrop in shape and comes in shades of

These strange structures glued to the leaf of a native plant resemble tiny seashells.

California red scale insects (*Aonidiella aurantii*) cover their tiny bodies with a leathery reddish-brown plate. Source: D. Papacek.

red, yellow, black, brown or white, depending on the species. The armour is tightly anchored to the plant surface, but can be prised off to reveal the tiny insect beneath. You can often find ants lurking around colonies of scale insects, feasting on the sugary honeydew they excrete a result of their sap-feeding diet. The ants chase parasitoids and insect predators and thus protect their food supply.

Rounded lumps or cushiony bumps on stems and foliage
Cushiony lumps on plants may be the bodies of soft scale insects, closely related to hard or armoured scales. The waxy covering of soft scales is actually a hardening of their external tissue, forming an important and inseparable part of their bodies. Unlike hard scales, this waxy covering cannot be removed without harming the insect. Soft scales can be green, yellow, brown, black, pink or white and come in a variety of textures. They feed on the sap of plants and secrete sugary honeydew, which can encourage ants and sooty mould. Females are fat and sluggish and the tiny mobile young are known as crawlers. Male scale insects are sometimes rare. Females can produce young without mating.

Cottony filaments on stems and leaves
Tiny balls of white fluff on plants are insects known as mealybugs, a close relative of scale insects (see above). Like scale insects, they use their needle-like mouthparts to feed on plant sap, excreting honeydew as a waste product. But while scale insects remain anchored to a plant, mealybugs possess legs and can actively crawl around. Mealybugs are covered in a white waxy substance, often extending into long tail-like filaments. Their offspring are tiny and pink and crawl about the plant looking for

A hard pink wax covers the delicate body of the female pink wax scale (*Ceroplastes rubens*). Source: D. Papacek.

CLEVER CLUES – THE STRANGE STRUCTURES AND EVIDENCE THAT INSECTS LEAVE BEHIND 153

Solenopsis mealybugs (*Phenacoccus solenopsis*) are pests on a variety of plants, including ornamentals, vegetable crops and cotton. This photograph shows the large adult females surrounded by multitudes of tiny offspring. Source: D. Papacek.

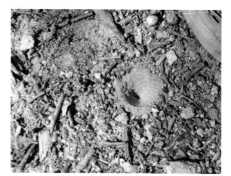

These funnel-shaped holes can be found in loose soil, such as the dirt underneath our houses.

Antlion lacewing larvae (*Myrmeleon* sp.) use their shovel-shaped head to dig pits in soil in order to trap ants.

a feeding site. The adult females are fairly sluggish and wingless but males have wings, look like tiny flies and can fly weakly.

Insects that may look like cottony filaments on stems and leaves

    Cottony cushion scale, p. 284
    Mealybug ladybird beetle larvae, p. 192

Can't find it here? The lumps or bumps you have found may be a nest or hideout (refer to next section) or eggs/egg cases (see p. 155).

Nests and hideouts

Many insects are master architects and construct elaborate structures in which to live, rear their young or help capture their food.

Funnel-shaped holes in sand or soil

Tiny funnel-shaped holes in sand or soil are usually the lair of an antlion, the larva of an antlion lacewing. The antlion larva digs the pit, dragging its abdomen backwards through the soft soil to form the hole and shovelling away dirt with its head. Once completed, it buries its body at the bottom and waits with the tips of its long pointed jaws poised to strike. Unsuspecting ants stumble into the trap and the antlion uses its head to flick sand at any that try to scale the walls to escape. Once seized, the ants are quickly dragged underground and sucked dry. The empty husk is tossed out of the pit at the end of the meal. You can find these pits in areas of undisturbed soil and sand, such as the dirt underneath your house.

Mounds of loose soil on lawns

Small mounds of fine dirt that resemble miniature volcanoes on lawns are the handiwork of a tiny yellowish-brown ant, known as the pasture funnel ant. The ants are rarely observed above ground, but their mounds are highly visible against the green grass of lawns, golf courses and footpaths.

Looking like miniature volcanoes, these piles of soil appear on our lawn after rain or soon after mowing.

A pasture funnel ant (*Aphaenogaster pythia*) moving dirt from the tunnels within its nest. Source: J. Dorey.

You can find frothy bubbles that look like spit, clinging to the stems of small trees and shrubs in your garden.

When it rains or when you mow your lawn, dirt and debris get knocked down into the tunnels of the nests. The ants work frantically to clear out this mess, depositing the dirt in a neat pile around the entrance.

Frothy bubbles on stems

Sometimes the branches of small shrubs are covered in white frothy blobs of bubbles. These are caused by the aptly named 'spittle bug', a small insect belonging to the Order Hemiptera that resembles a tiny cicada as an adult. The nymphs of these insects surround themselves in a frothy mass of sticky bubbles, often completely obscuring their bodies. It is not spit they surround themselves in, but a delightful combination of their bodily waste mixed with air bubbles. This frothy fortress protects the insect from dehydration and can deter predators and parasites.

Clumps of tiny brown 'vases' or big balls of dried mud on walls and brickwork

Mud dauber wasps and potter wasps (also known as mason wasps) are master builders and their carefully crafted vessels can be seen along walls, under eaves, tucked into brickwork and wedged into other sheltered areas.

Female mud dauber wasps are often seen at the muddy edges of puddles, gathering mud in their jaws, which they use to build and shape their nests. Each cell consists of a smooth vase-shaped vessel, which upon completion is stocked with an egg and several paralysed spiders (stung by the wasp) as food for the hatchling larva. Potter wasps are much larger and stockier than the

CLEVER CLUES – THE STRANGE STRUCTURES AND EVIDENCE THAT INSECTS LEAVE BEHIND 155

These small mud 'vases' can be found fastened to walls and around brickwork. Source: L. Woodmore.

Dome-backed spiny ants (Polyrachis sp.) incorporate pieces of vegetation into silk to construct their nests.

Mud dauber wasps (Sceliphron sp.) construct vase-shaped nests from muddy soil. Source: V. Dunis.

slender-bodied mud daubers and provision their nests with the caterpillars of butterflies and moths, rather than with spiders. Their nest consists of many individual mud cells, stacked neatly into a large ball.

There are many different species of mud dauber and potter wasps, each with its own method of nest construction. Some species build a single layer of vessels whereas others stack them in multiple layers, often covering the whole structure in an extra layer of mud for stability and protection.

Balls of leaves woven together in foliage
Several species of ants in Australia nest high in the foliage of trees, rather than underground. Their nests are constructed by weaving together the broad leaves of trees and shrubs into a ball-shaped structure, using silk. Adult ants do not produce silk (indeed, most adult insects cannot do so) and so enlist the help of ant larvae in building their nests. The maggot-like grubs are gently grasped in the jaws of the adult ants and dabbed against the surfaces of leaves, leaving behind a coating of sticky silk. Other ants drag the edges of the leaves together and hold them in place until the weaving is complete.

Can't find it here? The 'nest' you have found may actually be an egg case or cocoon. Refer to the next section.

Cocoons, cases and eggs

Many insects produce protective structures, using a variety of materials. These coverings may be used to bundle together eggs, or to provide protection during other stages of the insect's lifecycle.

Long 'cocoons' covered in sticks or leaves
Birds love feasting on fat juicy caterpillars. To protect themselves, case moth caterpillars weave 'sleeping bags' out of very strong silk, some of which are reinforced with sticks or leaves. With just their head

Long cases of leaves and sticks are often found attached to the trunks, stems and leaves of a variety of plants.

By popping its head and legs through an opening at the top of its bag, this Saunder's case moth caterpillar (*Metura elongata*) goes in search of suitable leaves to eat. Source: J. Dorey.

and thorax poking out of the top, these caterpillars can munch on leaves while remaining inside their protective fortress. We find these cases attached to the trunks, stems and leaves of many different types of plants, including natives, ornamentals and fruit trees. The caterpillar turns into an adult moth within the bag. The winged males fly away through the bottom of the bag, in search of wingless females which remain in the bag.

Shiny gold or silver globes dangling from shrubs

Shiny gold or silver globes can be found dangling from fig trees and oleander bushes in our gardens. While they look like miniature Christmas decorations, they are actually the chrysalis or *pupa* of the common crow butterfly. It is within this structure that the caterpillar transforms into an adult butterfly. There are several theories as to why the chrysalis comes in such a dazzling shiny form. Some scientists believe it is to advertise that there are toxic or distasteful compounds within its tissues. Another theory is that, up close, the reflective surface would act like a mirror, perhaps to startle potential predators.

A pale 'marshmallow' on twigs, shrubs or the ground

Praying mantids lay their eggs into a large egg case known as an ootheca, which

These shiny gold or silver globes are around 25 mm long and can be found on fig trees and oleander bushes.

Common crow butterflies (*Euploea core*) emerge from shiny gold or silver chrysalises. Source: E. Siegel.

A praying mantis ootheca deposited on the ground.

resembles a pale brown marshmallow. These can be found glued to the stems and leaves of plants, on the trunks of trees, or under rocks and leaf litter on the ground. To construct the ootheca, the female produces a special liquid within her body and uses the finger-like appendages at the end of her abdomen to spin it into a spongy foam. Between 10 and 400 eggs are encased in this frothy material, depending on the species of praying mantis that produces it. The foam eventually hardens, forming an insulating barrier that protects against extreme environmental conditions and some predators. It can sometimes be hard to determine if an ootheca is empty or not. While some have tiny exit holes in their walls, these are usually attributed not to the hatchling mantids but to the emergence of tiny parasitic wasps that have attacked the eggs within.

Tiny brown pellets in cupboards or attached to the surface of leaves and bark

In our homes, we sometimes come across hard brown shiny pellets that, upon closer inspection, resemble a tiny clutch purse. These are the egg cases of household cockroaches and are often found in cupboards, stuck to shelves or scattered on benches and floors. In the wild, they can be found among leaf litter, or glued to the leaves or bark of trees. Like praying mantids (see above), cockroaches lay multiple eggs into a special bag known as an ootheca. Each contains between 12 and 40 eggs, depending on the species, and females can produce many egg cases throughout their adult lives. Rather than depositing them on surfaces, some cockroaches prefer to carry around their ootheca protruding from the end of their abdomen.

These small oval-shaped pellets can be found on surfaces in our home or glued to leaves and bark. Source: K. Ellingsen.

Neat rows of hairs, evenly spaced on solid surfaces

Neat rows of delicate hairs, evenly spaced along solid surfaces, are the eggs of lacewings. They can be found fastened to the eaves of houses, tree trunks and the underside of leaves. When laying her eggs, the female lacewing touches a sticky drop of fluid from the end of her abdomen onto the substrate then stretches it out into a long strand. This stalk hardens and an egg is placed at the tip, out of reach of hungry predators such as ants. These strands are produced in batches, laid either in extended rows (green lacewings, see p. 92), or curved into the shape of a horse-shoe (blue eyes lacewing, see p. 214).

Tiny green lacewing larvae (*Mallada signatus*) hatch from their eggs. Source: J. Dorey.

Can't find it here? The structure you have found may be a nest or hideout. Refer to the previous section.

6
Insect Orders

How to use this chapter

Insects are sorted into different taxonomic groups known as Orders (see Chapter 2, p. 12). This chapter contains detailed information on the most commonly encountered insect Orders in Australia, including notes on their appearance, biology and lifecycles, as well as colour

Table 6.1: Information found in summary boxes

Feature	Description
Common name	The name or names members of the Order are commonly known as.
Order name	The proper scientific name for the Order.
Meaning of Order name	The names of insect Orders originate from Greek or Latin words and this translation reveals their meanings in English.
Number of species	The total number of species formally described within the Order, both in Australia and worldwide. These figures vary between sources and are not definitive. Some authors quote the **estimated** total number of species, a value that is inevitably much larger than the number of **described** species, as many insects still await discovery. Worldwide species numbers are taken from *Insect Biodiversity: Science and Society* by R.G. Foottit and P.H. Adler (Eds) and Australian species numbers from *A Field Guide to Australian Insects* by P. Zborowski and R. Storey (see Bibliography).
Key features	A summary of the main diagnostic features of adult insects within the Order.
Suborders	Some Orders are further divided into Suborders, a taxonomic group that generally consists of a group of closely related Families. While Suborders exist within many insect Orders, in this book they are only listed if they represent distinct groups easily recognisable to the average person. For example, many people would recognise crickets and grasshoppers (which belong in separate Suborders) as two different kinds of insects, and so the features of both Suborders have been listed in the summary box. Within this chapter, Suborders are dealt with separately only if they represent groups of insects that differ greatly in their biology. For example, aphids, cicadas and water bugs represent Suborders of the Order Hemiptera that differ so greatly in appearance and lifestyles that each has its own section in this book.
Synonyms	A list of any alternative names for the Order. These may be past classifications that are no longer valid or new names that, despite being currently valid, are not yet widely used in insect books and websites (see Chapter 2, p. 19 for more information).
Similar Orders	Pictures and information on other Orders whose members are similar in appearance or occupy the same kinds of habitats. This allows you to compare similarities and differences between Orders and will help you to rectify misidentifications.

photographs, further reading lists and suggestions of keywords for internet searches.

At the start of each Order, you will find a summary box that acts as a quick reference guide (see Table 6.1).

Bees, wasps, ants and sawflies – Order Hymenoptera

Summary – Bees, wasps, ants and sawflies

Order Hymenoptera, meaning 'membrane wing'

144 695 species worldwide, 14 871 species in Australia

Key features

- A narrow 'waist' between the thorax and abdomen, with the exception of sawflies (which have a uniformly broad body)
- Two pairs of membranous wings that are usually hooked together; wings folded flat over abdomen or held alongside body at rest
- Fore wings larger than hind wings
- Large compound eyes

Sawflies, Suborder Symphyta

Key features
- No constriction between thorax and abdomen
- Females with a saw-like ovipositor for laying eggs, although this is hard to see

Bees, wasps and ants, Suborder Apocrita

Key features
- A narrow constriction between the thorax and abdomen (occasionally hard to discern)
- Females with a needle-like ovipositor for egg-laying and sometimes stinging

The information pages for each group (bees, wasps, ants and sawflies) have information on other insect Orders you may have them confused with.

Steelblue sawfly (*Perga dorsalis*). Source: L. Woodmore.

Yellow paper wasp (*Polistes dominulus*). Source: L. Woodmore.

Description

The Order Hymenoptera is very large and diverse. It includes bees, wasps, ants and sawflies (which are not actually flies, but a type of primitive wasp). Members of this Order range from brightly coloured wasps as long as your thumb, to tiny ants only a few millimetres long. Some are accomplished fliers, others spend much of their life underground or living as larvae within the bodies of other insects.

Bees, wasps and ants differ from other insects by having a narrow 'waist', that is, the point where their thorax and abdomen meet is often greatly constricted. To be anatomically correct, it is actually the **second** segment of the abdomen that is narrowed – the very first part is broadly attached to the thorax. Referred to as the petiole, this narrow second segment increases flexibility of the abdomen for egg-laying and stinging. This constriction of the abdomen is absent in the more primitive sawflies.

Adult Hymenoptera have two pairs of transparent wings, with the fore wings larger than the hind wings. The fore and hind wings often lock together before flight through a series of hooks (called hamuli) on the edge of the hind wings. This increases the surface area of the wing, making flying easier. Only some types of Hymenoptera, such as worker ants and some species of wasps, remain wingless for their entire life.

Bees, wasps, ants and sawflies have large compound eyes and often three simple eye spots (ocelli) on the crown of their head. Their antennae come in a variety of shapes and sizes and sometimes differ greatly in appearance between males and females. A strong pair of biting jaws (mandibles) is used for killing and handling prey, for constructing nests and sometimes in defence. At the bottom end of the body is a long thin ovipositor – a modified egg-laying tube seen only in females. It can be blade- or saw-like to assist with laying eggs into plants, or sharpened into a hypodermic syringe for injecting venom.

The Order Hymenoptera is divided into two separate Suborders. Sawflies belong to Symphyta and are a primitive form of wasp. Bees, wasps and ants are all members of the Apocrita, the larger and more diverse of the two Suborders. As bees, wasps, ants and sawflies differ wildly in appearance and lifestyles, each has been dealt with separately, starting with the more commonly encountered bees, wasps and ants and lastly the more obscure sawflies.

Bees

Summary – Bees

Order Hymenoptera, Suborder Apocrita, Superfamily Apoidea*

16 333 species worldwide, 1652 species in Australia

Key features

- Narrow 'waist' between thorax and abdomen, although sometimes hard to discern
- Often with furry-looking bodies

- Two pairs of membranous wings that are usually hooked together; wings folded flat over abdomen or held alongside body at rest
- Fore wings larger than hind wings
- Large compound eyes
- Elbowed antennae

Honey bee (*Apis mellifera*).

*The Superfamily Apoidea also includes a few groups of wasps (not listed here) which are closely related to bees.

Don't confuse bees with:

Flies

Order Diptera, see p. 225

Things in common

- Often similar in shape and size
- Membranous wings that are folded flat over abdomen or held alongside body at rest
- Large compound eyes

Wasp-mimicking hover fly (*Mesembrius* sp.).

Differences

- The bodies of flies may be covered in bristle-like hairs, but are generally not furry in appearance like bees
- Flies have only one pair of wings, whereas bees have two pairs
- The antennae of flies are much shorter and thinner than those of bees

Description

When we think about bees, it is generally the orange and brown honey bee that springs to mind. However, the honey bee was introduced to Australia from Europe in the 1800s and they differ greatly in both appearance and behaviour from our native bee species.

Bees evolved from wasps and have a unique appearance and lifestyle. They are diverse in size and colour, from tiny black stingless bees that are quite shiny and only a few millimetres long, through to furry metallic green carpenter bees that are almost 20 mm long.

Like wasps and ants, bees have a narrow 'waist' between their thorax and abdomen. However, their bodies are shorter, stouter and furrier and their large rounded abdomens often make this constriction difficult to see. Bees are unique in that they have a series of branching hairs covering their bodies, making them look furry. Bees presumably evolved this special hair structure (which is unique among insects) to help them collect pollen more effectively. Wasps generally look much smoother and, although some have hairy bodies, these hairs are slender and not branched.

Bees have large compound eyes and three ocelli, which gives them excellent vision. A long tongue, or *proboscis*, enables them to lap up nectar from flowers and a pair of sharp mandibles aid in nest construction. Their two pairs of membranous wings are hooked together during flight and the fore wing is larger than the hind wing.

Diet and habitat

With a few exceptions, most native bees in Australia are solitary, with each female building her own nest and provisioning it with food for her young. While some solitary bees may place their nests close to those of other females, they do not cooperate to construct nests. This is the bee equivalent of owning your own home within a housing estate.

The nest of a solitary bee usually consists of a burrow, which is made up of individual cells for the young. Bluebanded bees excavate soil to make mud nests, whereas carpenter bees hollow out decayed wood and the stems of plants for their nurseries. Some bees construct resin nests in existing cavities, such as meter boxes and beneath window sills. Leafcutting bees line their cells with leaves that they carefully excise from plants – the perfect circles they cut are often evident in your garden. Mason bees, as the name suggests, partition off their cells with a thick mixture of mud.

Bees also differ from wasps in the food they provide their young. While many wasps hunt insects and spiders for their offspring, bees provision their nests with pollen and nectar. A variety of native and introduced flowering plants are targeted, although eucalypts and tea trees are especially important as food sources for bees throughout Australia.

The hairy bodies of bees allow them to effectively remove pollen from flowers and carry it back to their nests. Special brushes on their mid and hind legs help to groom pollen from their body then transfer it to the mouth, where it is mixed with saliva to make it sticky. It is then moved to special stout hairs on the bee's body known as the scopa, for transportation back to the nest. In some bees, these hairs form a pollen basket on the hind legs – a hollowed-out area surrounded by stiff hairs. You can see the big yellow balls of pollen on the legs of stingless bees and honey bees as they approach their hive. Other types of bees carry pollen on the hairy undersides of their abdomen.

Lifecycle

At certain times of the year, male honey bees form large swarms in the air, waiting for newly emerged virgin females to arrive. A female honey bee may mate with up to 20 different males in one of these swarms. For male honey bees, mating is lethal. After the deed is done, the male separates from the female, tearing off his nether regions in the process, killing him. His reproductive organs remain lodged within the female's body, until they are removed by either the female bee or another eager male.

Honey bees are social insects that live in hives made up of thousands of sterile females, ruled by a single fertile queen. Within a honey bee hive, the workers construct a labyrinth of cells, made from a pliable wax secreted from their bodies. After each cell has been stocked with honey, the queen bee deposits a single egg into each and the workers seal it off.

The rearing of young is quite different in solitary bees as **all** females are capable of laying eggs and each female is responsible for constructing and provisioning her own nest. In the warmer months, adult bees emerge from their nest and take to the air in search of a mate. The males are usually short-lived and die soon after mating. The female stores sperm in a special reservoir within her body called a spermatheca, until she is ready to fertilise her eggs.

After mating, the female solitary bee sets about building and provisioning her nest. Once an individual cell is completed and stocked with nectar and pollen, the female deposits a single egg and seals off the entrance, before repeating the process. The female usually dies before her offspring emerge as adults.

Bees, like all Hymenoptera, undergo complete metamorphosis, with a distinct egg, larva, pupa and adult stage. The maggot-like bee larvae are legless and usually pale. Like human infants, they rely entirely on their mother to provide them with adequate food and shelter. They undergo several moults before turning into a pupa, eventually emerging as a winged adult.

Defence

Bees have many enemies. They are a favourite food of predatory insects such as dragonflies, robber flies, praying mantids and assassin bugs. Well-camouflaged flower spiders lurk among the petals of flowers and pounce on thirsty bees as they land. Many species of birds also feast on bees, among which bee-eaters are notorious.

We all know that bees sting, but not all types do so. Australia is home to several species of tiny black stingless bees. For defence they use their tiny jaws to latch onto enemies and bite them. Other native bees can sting repeatedly, however they are generally shy and sting only if handled or provoked.

The honey bee is the only bee in Australia that leaves its stinger behind in the body of the victim. The stinger is barbed, becoming snared in the soft flesh of our skin. When the bee flies away, the stinger (and a large portion of its abdomen) is torn off. Complete with a sac of venom, this organ continues to pulse and wriggle, pumping more painful toxins into your system. Always use a fingernail to scrape your skin where you have been stung, to ensure the stinger is completely removed.

In the 1980s a US entomologist, Justin Schmidt, and colleagues developed a pain scale index for stinging insects such as bees, wasps and ants. Species were given a ranking of zero to four: four is the most painful and zero is used for insects with stingers that are unable to pierce human skin. On this index the honey bee received a rating of two out of four, with the duration of pain lasting up to 10 minutes. Schmidt likened the experience to 'a match head that flips off and burns on your skin'. While the 'ouch' factor is relatively low, care must be taken with bee stings as allergic reactions can be deadly.

Goodie or baddie?

Bees might have a bad reputation to some people because their sting is painful. However, as far as insects go, no other group has such a profoundly beneficial impact on our way of life.

We all know that bees produce honey, but it does not stop there. Bees are vital pollinators of our trees, shrubs, pastures and fruit and vegetable crops. If fact, the production of nearly one-third of all the food we eat relies on the pollinating behaviour of insects such as bees. Albert Einstein understood the value of bees and has been famously linked to the quote, 'If the bee disappeared off the surface of the globe, then man would only have four years of life left. No more pollination, no more plants, no more animals, no more man.'

Future Fido?

Bees can make surprisingly good pets, providing you choose the right type! Native stingless bees are a social bee that is easily domesticated and can be kept in small hives in your garden. They produce delicious honey known as sugarbag and provide a valuable pollinating service.

Fascinating facts

- Cuckoo bees are very cunning. Instead of building their own nest, they lay their eggs in the nests of other bee species. The hatchling cuckoo bee larva has strong sharp jaws, allowing it to swiftly destroy the resident eggs and larvae within its stolen home.
- Let's hear it for the boys! Only female bees (and wasps and ants) are capable of stinging. The stinger is made from a modified tube for laying eggs. Males don't lay eggs and so lack this organ.
- Honey bees move quickly between flowers, efficiently gathering nectar and pollen before moving to the next one. An average worker bee reportedly visits up to 1000 blossoms each time it leaves the hive to forage.
- In Japan, giant hornets raid bee hives and kill the occupants. By working together, the tiny bees can fight back against these much larger wasps. When a hornet enters the hive, the workers swarm around it, pressing together their bodies and gently buzzing their wings. This causes the wasp's body to overheat, eventually killing it.
- Bad news regarding honey bee stings. When a bee stings you, she releases a chemical signal known as an alarm pheromone, which can be detected by other bees in the vicinity. Basically, this pheromone is a bee's way of saying 'Hey guys, I've had to sting someone – come over here and help me!' Other bees can lock on to this chemical target and may sting you as well.

Learn more

In this book:

Bluebanded bees, p. 226
Carder bees, p. 199
Honey bees, p. 286
Larvae, p. 69
Leafcutting bees, pp. 147–148
Megachile monstrosa, p. 173

In other books:

Grissell E (2010) *Bees, Wasps and Ants: The Indispensable Role of Hymenoptera in Gardens.* Timber Press, Oregon.
Detailed information on the role of bees, wasps, ants and sawflies in our

Some bees, like this neon cuckoo bee (*Thyreus nitidulus*), have bright green or blue and black bodies. Source: E. Siegel.

environment. Features great colour photographs and is an excellent book for gardeners. ☾

An explanation of the above book ranking system can be found in Further Reading.

Online, type in:

Colony collapse disorder – learn about the alarming phenomenon in which worker honey bees disappear without a trace.

Schmidt pain index – for more details on the most painful insect stings, including bees.

Varroa mite – learn more about this parasitic mite that is threatening commercial honey bee hives around the world.

Stingless bees (*Tetragonula carbonaria*) gathering pollen from a water lily.

Resin bees (*Megachile* sp.) are solitary bees that build their nests in pre-existing holes in timber, sealing the entrance with a mixture of chewed-up plant fibres and resin. Source: L. Woodmore.

Wasps

Summary – Wasps

Order Hymenoptera, Suborder Apocrita

107 875 species worldwide, almost 10 000 species in Australia

Key features

- Narrow 'waist' between thorax and abdomen, usually long slender bodies
- Two pairs of membranous wings that are usually hooked together; wings folded flat over abdomen or held alongside body at rest
- Fore wings larger than hind wings
- Large compound eyes
- Antennae long and thin

Yellow-banded ichneumon (*Xanthopimpla rhopaloceros*). Source: K. Ellingsen.

Don't confuse wasps with:

Flies

Order Diptera, see p. 225

Things in common

- Often similar in shape and size
- Membranous wings that are folded flat over abdomen or held alongside body at rest
- Large compound eyes

Differences

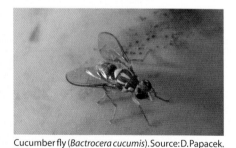

Cucumber fly (*Bactrocera cucumis*). Source: D. Papacek.

- Flies have only one pair of wings, whereas wasps have two pairs
- The antennae of flies are much shorter and thinner than those of wasps

Description

Wasps are easily recognised by the narrow constriction where their thorax and abdomen meet. This tiny 'waist' makes wasps extremely flexible, allowing them to manoeuvre the tip of their abdomen with ease, either to deposit eggs with pinpoint accuracy or plunge a stinger into an unfortunate victim.

The body of a wasp is slender and sleek, compared to their furry, frumpy bee cousins. A pair of highly sensitive antennae, large compound eyes and three ocelli help them to navigate and sense their surroundings. Wasps have two pairs of transparent wings, which may be clear, yellow, brown or smoky black. The fore wing is larger than the hind wing and the wings are usually joined together with hooks during flight.

Wasps range enormously in size and shape. Yellow and black spider-hunting wasps can grow bigger than your thumb and are equipped with powerful wings for flying. Cuckoo wasps are tiny and come in beautiful shades of metallic blue and green. Some minute parasitic wasps, even as adults, are no bigger than the full-stop at the end of this sentence. If you wanted to see one, you would have to use a microscope!

Diet and habitat

Wasps are usually active during the day and prefer warm terrestrial environments such as gardens, flowering grasslands and forests, where adults feed mostly on nectar. However, when it comes to the living arrangements and diet of their young, wasps use such a diverse range of habitats, food types and lifestyles that it is necessary to split them into several different groups.

Gall-forming wasps

Some wasps begin their lives as parasites within the tissue of plants. Certain species of tiny wasps (known as chalcidoid wasps) lay their eggs in the stems, leaves, seeds and flower buds of native plants such as eucalypts and wattles. The feeding action of the wasp larvae causes the plant to produce layers of extra cells, eventually forming a swelling of plant tissue (a gall) around the larva. The larva hides inside this igloo-like home, feeding on the plant tissue within it.

Parasitic wasps

Parasitic wasps invade the bodies of other insects, feasting on their internal tissues. The female wasp inserts her eggs into the host using her needle-like ovipositor. In some wasps, this needle can be 10 times longer than the rest of their body, allowing them to probe deep into soil, plant stems or wood in search of suitable hosts. Other species glue their eggs to the outside of the host's body or scatter them over the leaves upon which their host feeds, so they are ingested. From these eggs hatch maggot-like larvae, which burrow through the body of the host, eventually killing it.

Almost every insect imaginable, and many types of spiders, are victims of at least one species of parasitic wasp. Many wasps target eggs, most commonly those of true bugs, butterflies and moths. Some wasps even swim underwater in search of the aquatic-dwelling eggs of dragonflies, damselflies and diving beetles. Wasps also parasitise the nymphs of grasshoppers and aphids, the larvae of sawflies and moths, and adult spiders.

Predatory wasps

Not all wasps live together in nests. In fact, most species are solitary, with individual females making their own nest and provisioning it with food for their young. Some females may place their nests close to those of other females, much like humans building homes within a housing estate.

Female predatory wasps often construct their nests from mud, or dig tunnels in the soil. Once the nest is completed, the wasp goes in search of insects or spiders to stock it. A sting is used to paralyse her prey, which is then carried back to the nest and tucked neatly into the cell. Lastly, the female lays an egg on or beside the body of the paralysed host, which remains immobile but fresh.

The maggot-like larva feeds on its provisions until it reaches adulthood.

Social wasps

The most highly evolved lifestyle is that of the social wasps, where many individuals of the same species live and work together in a communal nest or *colony*. There is a distinct hierarchy and each individual cooperates to ensure the success of the colony. There is a single queen and many hundreds or thousands of infertile female workers. The queen single-handedly controls the activities within the colony and is solely responsible for all egg-laying. The queen wasp keeps her minions in line through physical aggression, bullying her workers by head-butting them and pushing them around, or through the secretion of chemicals that influence the behaviour of the other wasps. The result of this harassment is that other females in the nest fail to develop their ovaries and are unable to lay eggs.

Social wasps progressively feed their young throughout their development rather than stockpiling paralysed food as the predatory wasps do. Insects such as caterpillars are killed, chewed up and swallowed by foraging workers. Upon returning to the nest, the wasps regurgitate this meal for the larvae, much like a bird tending to its chicks.

Lifecycle

All wasps develop in much the same way, despite their differences in feeding types and habitat. They undergo complete metamorphosis, with an egg, larva, pupa and adult stage.

The larvae are quite variable in form but are usually legless, pale and grub-like. Most are unable to move from the location where they hatch, and so rely entirely on their mothers to lay the egg in a suitable place with enough food for them to complete their development.

The formation of a pupa usually takes place within a cocoon woven from silk. It may be found on or near the body of the host, or it sometimes dangles nearby on a strand of silk. Other individuals construct a papery cocoon in soil or leaf litter.

Some parasitic wasps practise polyembryony – a process that can be likened to the development of identical twins in humans. Encyrtid wasps lay a single tiny egg into the egg of a moth. Once the moth egg has hatched into a caterpillar, the parasitoid egg divides (like a human egg that yields identical twins) but it does so numerous times – up to a thousand embryos develop. Of the larvae that emerge, some act like assassins, single-mindedly seeking out and destroying the larvae of any other parasitic insects developing within the caterpillar. Eventually these individuals die, making way for their siblings to feed and grow within the host's body.

Defence

Female wasps are equipped with a sharp stinger capable of delivering a powerful dose of venom. Wasps often boast bright warning colouration – striking combinations of orange or yellow markings, contrasting brightly against brown or black backgrounds. These bright colours send a warning to enemies that the wasp is a dangerous insect and not to be messed with.

Unlike honey bees, which have a barbed stinger that becomes lodged in the flesh of their victim, wasps have a smooth sharp stinger which glides in and out of tissue quickly and cleanly. This means that a single wasp is capable of delivering multiple stings. The venom can cause extreme pain and localised swelling and can cause life-threatening allergic reactions in a small percentage of humans.

On the Schmidt pain index (see p. 166) some wasps rank very highly in terms of their 'ouch' factor. A paper wasp, for example, is given a value of three, with the pain lasting anywhere from five to 15 minutes. Schmidt likens the sensation to 'spilling a beaker of hydrochloric acid on a paper cut'.

Goodie or baddie?

Hornets, European wasps and paper wasps have a bad reputation, because they can deliver painful stings. They also have a habit of showing up unwanted in and around our homes and gardens, such as when paper wasps build nests on the eaves of houses and hornets buzz around when we eat outdoors.

Despite their bad reputation, wasps provide a vital service to humans. Many wasps are predators of other insects and their voracious appetites help to keep insect populations, including many pest species, at tolerably low levels. The highly selective nature of parasitic wasps makes them excellent candidates for use in biological control. Females attack only certain species of insects. This means they can be mass-released into crops and orchards to control pests, without harming beneficial insects such as bees (unlike most insecticides).

Fascinating facts

- 'Fairy fly' is the adorable name given to a tiny parasitic wasp that attacks the eggs of booklice. Adult males are around 0.14 mm long, making them one of the smallest flying insects in the world. To give you an idea of how small we are talking, you could line up around 57 fairy flies, head to tail, along a single grain of rice.

- Female flower wasps are wingless, an evolutionary modification to their bodies that makes digging through dirt in search of beetle larvae, on which they lay their eggs, much easier. In a move that strikes fear into the heart of feminists, these wasps are completely dependent on the male wasps for survival. The winged males grasp the females and fly them off to visit flowers, where they consume nectar and pollen. Once they have eaten their fill, they are promptly delivered back to their nest to resume their maternal duties.

- Talk about fussy eaters! Trigonalid wasps scatter their eggs on leaves where they are gobbled up by moth caterpillars. Upon hatching inside the host, the tiny wasp larvae go in search of food, but it is not the tissue of the caterpillar they are interested in. Instead, they are on the lookout for parasitic larvae of other insects (e.g. tachinid fly parasitoids) **already** inside the body of the caterpillar. If no such larvae are present, the trigonalid larva waits patiently until the caterpillar is parasitised – but it sometimes starves to death while waiting.

- Imagine being served dinner only to have it turn around and eat you! This can happen if you are the unlucky larva of a paper wasp. The eggs of a certain type of trigonalid wasp are eaten by moth caterpillars, which are hunted by paper wasps. The wasps return to their nest and regurgitate the chewed-up caterpillars for their hungry larvae. Within this pile of mushy caterpillar parts, the trigonalid larva hatches then proceeds to eat the unsuspecting paper wasp larva.

- Spider wasps, as the name suggests, hunt arachnids as a food source for

This colourful wasp (Family Eulophidae) is tiny in comparison to the eye of a needle. Source: J. Dorey.

their larvae. But when it comes to the trouble of building their own nest, some of these wasps prefer to cut corners. Trapdoor spiders are lured out of their underground burrows and are stung by spider wasps. The now-incapacitated spider is unceremoniously shoved back into its home, along with a new tenant in the form of a wasp egg. The egg soon hatches into a larva and consumes its comatose roommate.

A gasteruptiid wasp (Family Gasteruptiidae, on the left) closely observes a native bee (*Megachile monstrosa*) constructing her nest. If given the chance, this parasitoid wasp will use her slender abdomen to deposit her own eggs into the nest of the bee. Source: L. Woodmore.

Learn more

In this book:

Blue-ants (actually a wasp), p. 130
Mud dauber wasps, pp. 154–155
Orchid dupes, p. 129
Potter wasps, pp. 154–155
Yellow paper wasps, p. 162

In other books:

Grissell E (2010) *Bees, Wasps and Ants: The Indispensable Role of Hymenoptera in Gardens*. Timber Press, Oregon.

(a) Spider wasps (Family Pompilidae) use their speed and powerful venom to hunt down and paralyse spiders as food for their young. Source: K. Ellingsen.
(b) These native paper wasps (*Polistes* sp.) are social insects that build large papery nests made from chewed plant material in which to house their young. Source: K. Hiller.

Detailed information on the role of bees, wasps, ants and sawflies in our environment. Features great colour photographs and is an excellent book for gardeners. ✱
An explanation of the above book ranking system can be found in Further Reading.

Online, type in:

Citrus gall wasp – the larvae of these tiny wasps form woody galls on citrus plants such as lemons and oranges.
Fig wasp pollination – learn about an extraordinary relationship between figs and the tiny wasps that pollinate them. Add 'video' to your search to view documentary clips showing the fig wasps in action.
Schmidt pain index – for more details on the most painful insect stings, including wasps.
Tarantula hunting wasp – hair-raising videos and images of female wasps hunting down spiders as a food source for their young.

Ants

Summary – Ants

Order Hymenoptera, Suborder Apocrita, Family Formicidae

11 946 species worldwide, 3000 species in Australia

Key features

- Small, usually brown or black
- A narrow 'waist' between the thorax and abdomen
- Wings present in reproductive ants, absent in soldier and worker ants
- Two pairs of membranous wings that are often hooked together and are folded flat over the abdomen at rest

Bulldog ant (*Myrmecia* sp.).

- Fore wings are larger than hind wings
- Large compound eyes
- Elbowed antennae

Don't confuse ants with:

Termites

Order Isoptera, see p. 265
Things in common

- Similar in size and appearance (termites are sometimes referred to as 'white ants')
- Both are social insects, living in colonies and divided into distinct castes

Differences

- Termites have a fat abdomen, i.e. the point where their thorax and abdomen joins is quite wide. In ants it is constricted, giving them a narrow waist (see Question 18 of the identification key (p. 48) for an illustration of this)
- The wings of termites (when present) are equal in size, whereas the fore wings of ants are larger than their hind wings
- Termites have straight antennae, ants have elbowed antennae

See the box on pp. 179–180 for other ant-mimicking organisms.

Giant northern termite (*Mastotermes darwiniensis*). Source: Science Image.

Description

Ants are highly evolved social insects belonging to a distinct Family within the Hymenoptera known as Formicidae. They range in size from tiny insects that are no more than a few millimetres long, to menacing bulldog ants which, at around 30 mm, are one of the largest ants in the world.

Ants, like bees and wasps, have a distinct 'waist', composed of a narrow constriction called the petiole and a swollen lower half that joins to the front of the abdomen, known as the gaster. An ant's body holds many tiny glands that produce special chemical signals known as pheromones, which are used in communication. A pair of highly sensitive, elbowed antennae enables ants to detect pheromones from other ants, as well as other chemicals from the surrounding environment.

All ants are social insects – many individuals of the same species live and work together in a colony. A communal nest houses adult ants and the brood, which is made up of eggs, larvae and pupae. A distinct hierarchy is evident among adult ants and each individual is allocated to a special type or *caste* within the colony, with its own role.

Workers – This group of ants is made up entirely of infertile wingless females. Workers, as the name suggests, are responsible for all the labour within a colony. There are foragers (skilled navigators that leave the nest in search of food), nurses (which care for the eggs and developing young) and builders (which are responsible for constructing and maintaining the nest). Like humans, worker ants may become eligible for career upgrades based on their age and skill level. For example, nurse ants that toil away in the brood chambers may, when they are older and more experienced, be promoted to foragers that are allowed to leave the confines of the nest to search for food.

Soldiers – Some species of ants include large burly individuals that take no part in nest-building or foraging. Rather, these infertile females dedicate their entire lives to protecting the colony and guarding the queen. These 'bouncers' of the ant world are usually equipped with strong mandibles and ferocious stings to help chase away intruders.

Queens – Virgin queens are produced within a colony at certain times of the year and fly from the nest to mate and establish new colonies. They have two pairs of wings and strong flight muscles. Their large abdomens are equipped with a well-developed set of ovaries and a large reserve of stored fat to assist with egg-laying. Compared to worker ants, queens have larger compound eyes and three simple eyes or *ocelli* to help them navigate through the air.

Winged males – Winged female ants are called queens but winged male ants are not referred to as kings, probably because they generally do not survive long after mating and play no role in building or controlling the nest. The winged males are smaller than the queens and also come equipped with large compound eyes and ocelli to help them navigate.

Diet and habitat

Ants are one of the most ubiquitous animals on our planet and, with the exception of extremely cold regions, can be found in all kinds of habitats. Ants are generally long-term tenants, with most species occupying a perennial nest excavated within soil or rotting wood. Some species use existing cavities, such as the hollowed-out stems of plants, crevices underneath rocks or between pavers, and wall cavities within our homes. A few species of adventurous ants build their nests on mangrove flats around northern Australia. Twice a day, the nest becomes inundated with tidal water and the ants are forced to shift to higher ground or retreat into air-filled pockets within the mud to avoid drowning. The advantage of occupying such an undesirable environment means there is very little competition from other ants for resources.

Adult ants generally feed on liquids, while their larvae are given solid foods. Many ants are fierce predators, attacking and killing prey many times their own size. Adult ants drink the juices from the prey's body and the larvae polish off the rest. Meat ants scavenge on the carcasses of other animals and can rapidly strip the flesh from the skeletons of small lizards and frogs. Other ant species collect nectar and sap from plants, or forage for seeds.

Ants have developed an interesting relationship with sap-sucking insects such as aphids and mealybugs that feed on the stems and leaves of plants. These insects excrete a sugary substance from their bodies, known as honeydew, and ants treat them like living vending machines. They loiter around the bugs, regularly getting their sugary fix, and in return fiercely guard them against predators. Some mealybugs live underground, feeding on the roots of plants. Ants herd them together and lead them to special enclosures within their own nests, where they are cared for and milked of their honeydew, like a miniature dairy farm. Tiny *Acropyga* ants are so dependent on their 'cattle' that a virgin queen carries an egg-producing mealybug with her as she leaves the colony in search of a mate. When she builds a nest of her own, she thus stocks it with honeydew-producing bugs.

Lifecycle

Once a year, usually on warm humid days after it has rained, thousands of winged ants leave the colony and take to the air in a nuptial flight. These flights are synchronised with neighbouring colonies of the same species, and presumably take place just after rain because it softens the hard soil and makes nest-building easier.

Once she has mated, the winged female searches for a suitable place to build a nest. Upon landing, she scrapes or bites off her wings, as these cumbersome appendages are not needed for a life underground. The young queen digs a small chamber and lays her first batch of eggs, which she then cares for. Once they are reared through to adults, her first generation of offspring is responsible for constructing a proper nest.

Unlike male termites, which enjoy a long life within a colony, male ants all die within a few days of mating. The queen ant stores sperm from the male in a special capsule known as a spermatheca attached to her reproductive system within her abdomen. Each egg is fertilised using this reservoir, meaning females can continue to produce viable eggs for many years after mating.

Ants undergo complete metamorphosis, hatching from their eggs as small, pale, grub-like larvae. They are fed and nurtured by the worker ants, much like human infants. If you have ever disturbed an ant nest, you may notice the workers rush to pick up the brood to carry them to safety. To make this easier, the eggs are often laid together in sticky clumps and the larvae have special hairs on their bodies that help them stick together, so they can be quickly carried off in little groups. The larvae eventually turn into pupae and emerge finally as adult ants, ready to join their legions of sisters working within the colony.

When food is limited, all individuals will emerge from their pupae as infertile female workers. However, in times of plenty, certain ant larvae are selected by the queen to receive extra nourishment. These individuals will grow into winged reproductive ants who will leave the colony to mate and establish new colonies.

Defence

Ants are tiny, and so are readily picked off by a variety of animals, most notably birds such as magpies, lizards such as thorny devils, and frogs, spiders and antlions. Ants wage war against rival ant and termite colonies and therefore have an arsenal of defensive strategies at the ready.

Ants rely on strength in numbers: if a lone ant runs into trouble, it rushes back to the nest to recruit an army of helpers. Soldier ants often have large jaws that can bite or pinch larger animals, or cleave off the heads, legs and antennae of rival ants and termites.

Ants also use chemical warfare to protect themselves and defend their nests. Like bees and wasps, many ants can inflict a painful sting. Multiple stings are common if you stray too close to an ant nest, as many ants often attack at once and each ant can sting repeatedly. Bulldog ants and jumping ants are some of the worst offenders in the Australian bush. These large, rather aggressive ants have a very painful sting and, in rare cases, severe allergic reactions have resulted in death. Despite being only 2 mm long, the introduced fire ant delivers a very painful sting for its size.

Some species of ants simply spray their toxins, rather than injecting them. Formic acid is used widely among ants and gives them a characteristic smell when crushed. Green tree ants first puncture the skin of their victim, then tilt their abdomen and blast a jet of acid into the wound, almost guaranteeing a hasty retreat.

Goodie or baddie?
Many people are not fond of ants. Ants invade our homes and raid our kitchens in search of sweet treats and crumbs. In our gardens, ants attack our seedlings and build nests in outdoor furniture and pavers. Pasture funnel ants pile up mounds of dirt around the entrance to their nests, making our lawns look unsightly. And who hasn't been stung on the back of the legs at least once by a greenhead ant while lounging around on the grass? In commercial crops, ants can be a serious pest, mostly because they protect honeydew-producing insects such as aphids and mealybugs, whose feeding can damage shoots and fruits and transmit plant diseases.

Despite all this, ants play a vital role in nature. The predatory lifestyle of some ant species makes them good for the biological control of insect pests, particularly in fruit orchards. Their endless nest-building helps in the formation, mixing and aeration of soil. Foraging ants collect seeds and store them underground, where they are safe from fire and scavengers, providing a vital dispersal service. Ants are also nature's street-sweepers, collecting scraps, carcasses and rubbish and taking it to their nests as food.

Fascinating facts
- The complex social hierarchy within an ant nest is controlled by pheromones deployed by the queen. These special chemical compounds 'brainwash' the inhabitants, so they move about doing the queen's bidding like drug-addled zombies!
- Kamikaze ants?! In Malaysia, there are ants that act like suicide bombers. Along the length of their bodies are large glands filled with toxic chemicals. When confronted with a hungry predator or an army of rivals, these ants suddenly contract the muscles in their abdomens, causing the glands to rupture and shower their foes with acid, killing themselves in the process.
- If you took all the ants in the world and popped them on the scales then did the same for all the humans, the weights would be roughly equal. Given an average ant is less than one-millionth our size, it means there are a lot more ants than people on our planet!
- An Australian ant, *Camponotus perthiana*, holds the record for the longest living ant – a queen survived in a laboratory nest for just over 23 years.
- On the Schmidt pain scale (see p. 166) one of the highest-ranking critters was the bullet ant, which received a ranking of four. Schmidt described the pain (which can last for up to 24 hours) as a 'pure, intense, brilliant pain. Like fire-walking over flaming charcoal with a 3-inch rusty nail in your heel'.

Thankfully, this ant lives in Central America and is not one you will find at your next picnic.

- Green tree ants (also referred to as weaver or citrus ants) are the earliest recorded example of insects being used in the biological control of pests. A passage of writing dating back to 304 CE describes bags of these ants being sold in marketplaces, to be transported to citrus orchards where they were used to control a variety of insect pests.

- Ever heard of social parasites? No, not those annoying people that latch onto us at parties! Some species of ants are too lazy (or cunning) to build their own nests. Instead, newly mated females sneak into the nests of other ant species and kill their queen. The new queen manipulates the resident workers into caring for her own eggs and young. Without their own queen laying eggs, the original species eventually dies out and the new ants take over the nest.

When is an ant not an ant?

Ant-mimicking spider (*Myrmarachne* sp.). A quick count of the legs confirms it is a spider (eight legs), not an ant (six legs). Source: P. Chew.

Myrmecomorphy is the phenomenon by which other arthropods mimic the appearance and behaviour of ants. It is beneficial to look like an ant. They possess a myriad of highly effective defensive tactics, such as powerful biting jaws, painful stings and corrosive formic acid, which means predators are often reluctant to approach them. When identifying insects in your home or garden, be on the lookout for these (and other) highly convincing ant-mimics.

Brown bean bug nymph (*Riptortus serripes*).

Ant-mimicking katydid nymph (*Torbia* sp.). Source: Hena Gecan, Attribution 2.0 Generic.

Newly hatched spiny leaf insect (*Extatosoma tiaratum*). Source: A. Hiller.

Learn more

In this book:

Dome-backed spiny ants, p. 155
Giant bull ants, p. 266
Greenhead ants, p. 131
Larvae, p. 68
Meat ants, p. 132
Pasture funnel ants, pp. 153–154
Sugar ants, p. 271

In other books:

Burwell C (2007) *Ants of Brisbane.* Queensland Museum, Australia.
A beautifully photographed, handy pocket guide to ants. As many of the ants in this book are also found in other areas of Australia, this book can be used for ants found in places other than Brisbane. ●

Clyne D (2010) *All About Ants,* New Holland, Sydney.
A great, easy-to-read guide to the wonderful world of ants. ●

Grissell E (2010) *Bees, Wasps and Ants: The Indispensable Role of Hymenoptera in Gardens.* Timber Press, Oregon.
Detailed information on the role of bees, wasps, ants and sawflies in our environment. Features great colour photographs and is an excellent book for gardeners. ●

Hölldoblet B, Wilson EO (1994) *Journey to the Ants: A Story of Scientific Exploration.* Belknap Press, Cambridge, MA.
This Pulitzer Prize-winning book contains interesting, cleverly written tales of life inside an ant nest. One of the best books out there if you want to learn more about ants. ●

Keller L, Gordon E (2009) *The Lives of Ants.* Oxford University Press, London.
An easy-to-understand book on the biology of ants. ● ●

An explanation of the above book ranking system can be found in Further Reading.

A foraging spider ant (*Leptomyrmex* sp.) carries a land hopper (Family Talitridae) back to its nest. Source: J. Dorey.

Golden-tailed spiny ants (*Polyrachis* sp.) feed on honeydew from a brown leafhopper nymph (*Alotartessus iambe*). Source: P. Chew.

The worker caste of coastal brown ants (*Pheidole megacephala*) come in two different sizes – minors and majors. The latter have greatly enlarged heads. Source: J. Dorey.

Online, type in:

Ant mimics – learn more about myrmecomorphy, the phenomenon by which many small creatures (e.g. spiders, katydid nymphs, true bugs) mimic ants in both appearance and behaviour.

Army ants – watch videos of these huge swarms of ants consuming everything in their path.

Exploding ant – these suicide bomber ants spray their foes with acid.

Giant ant hill excavated – watch a team of scientists unearth an enormous underground ant nest in Brazil by pouring concrete into holes and excavating the solidified tunnels.

Leafcutter ants – these resourceful ants make mulch gardens from leaves they cut from plants, and feed on the fungi they cultivate.

Schmidt pain index – for more details on the most painful insect stings, including ants.

Trap-jaw ant – this ant has the fastest bite of any animal.

Sawflies

Summary – Sawflies

Order Hymenoptera, Suborder Symphyta

8541 species worldwide, 176 species in Australia

Key features

- No constriction between thorax and abdomen
- Two pairs of membranous wings that are usually hooked together; wings folded flat over abdomen or held alongside body at rest
- Fore wings larger than hind wings
- Large compound eyes
- Antennae long and thin
- Females with a saw-like ovipositor for laying eggs (although this is hard to see)

Cattle-poisoning sawfly (*Lophyrotoma interrupta*).

Don't confuse sawflies with:

Flies

Order Diptera, see p. 225

Things in common
- Often similar in shape and size
- Membranous wings folded flat over abdomen or held alongside body at rest
- Large compound eyes

Differences
- Flies have only one pair of wings, whereas sawflies have two pairs
- Flies have antennae that are much shorter and thinner than those of sawflies

Black-headed hover fly (*Melangyna* sp.).

Description

Sawflies are not, as their name suggests, flies, but rather a small group of primitive wasps. Their stout bodies are often brightly coloured and can reach up to 55 mm in length. Sawflies differ greatly from other members of the Hymenoptera in both their appearance and lifestyle and are therefore placed in their own Suborder, known as Symphyta.

At a glance, sawflies do indeed resemble flies. They have broad, barrel-like bodies and lack the typical 'waist' seen in bees, wasps and ants. A closer examination reveals two pairs of membranous wings, unlike flies, which have only one pair. The fore wing is larger than the hind wing and has a more complex pattern of veins than those seen in bees or wasps.

The defining feature of sawflies is their serrated, saw-like ovipositor – a special organ at the end of the female's abdomen. Like a pocket knife, this blade opens out and is used to cut into plant tissue so that eggs can be deposited there. Unlike bees, wasps and ants, sawflies do not use their ovipositor in defence and cannot sting.

Diet and habitat

Adult sawflies generally feed on nectar, although a few groups prey on other insects. While we rarely come across the adults, their larvae and the damage they cause through feeding is sometimes very obvious.

The larvae of sawflies look and behave a lot like the caterpillars of butterflies and moths. They use their strong mandibles to feed on the leaves of plants. Some species skeletonise the leaves, feasting on the soft tissue and leaving only the tough veins behind. Other species strip the entire leaf, and may completely defoliate the tree. Often the larvae aggregate in large groups on leaves and branches by day and spread out among the foliage at night to feed.

Some larvae are reclusive, preferring to feed in concealed locations within the plant. Leaf miners (e.g. the leafblister sawfly) burrow between the upper and lower layers of the leaf, forming a small tunnel in which to shelter. Others tunnel their way through the stems of plants, or form tumour-like galls on leaves and stems.

Species in one very unusual group of sawflies, known as wood wasps, avoid plant material and are instead parasitoids of other insects. Female wood wasps lay their eggs in fallen logs, where wood-boring beetle larvae feed in tunnels within the dry wood. The wood wasp larvae not only devour the beetle larvae, but may also parasitise other kinds of stem-boring sawflies.

Lifecycle

Adult sawflies are usually short-lived and so finding a mate and laying eggs is their number one priority. Some species are remarkably good mothers (for an insect!) and display a great deal of care when it comes to protecting their eggs and young. The females use their ovipositor to cut slits in plant stems and wood and tuck their eggs safely inside, away from hungry predators. Others stand guard over their egg masses and young larvae, loudly buzzing their wings to scare away enemies.

Sawflies undergo complete metamorphosis, starting off life as a grub or *larva* before becoming a pupa, inside which they transform into a winged adult. However, their larvae are very different in appearance from the pale legless grubs of bees, wasps and ants. Young sawflies have long caterpillar-like bodies, with a strong head capsule and chewing mandibles. They have six stumpy legs located near their head and may also have a series of suckers or *prolegs* towards the end of their bodies to help them cling to plants.

After moulting their skin several times through their development, the larvae drop to the ground and form tiny silken or papery cocoons in the soil or leaf litter. Other, more cryptic feeders instead pupate inside their tunnels within the stems and leaves of plants.

Defence

As with most insects, adult sawflies are vulnerable to attack from a range of animals including spiders, birds, lizards and predatory insects. As they are quite harmless as adults, they must rely on various ways to deter enemies.

Some species stridulate – the process in which insects rub various parts of their bodies together to produce a startling sound. In the case of sawflies, they move a special spot on their fore wings over a patch of rough bumps called cenchri located on their thorax, to make an audible noise. Despite their lack of a stinger, some sawflies mimic the bright patterns and warning colouration of their dangerous wasp cousins – a nifty trick for deceiving enemies.

Sawfly larvae have their own tactics to avoid becoming an animal's lunch. Many use the notion of 'safety in numbers', aggregating in large clumps on the leaves and stems of plants. If disturbed, the group will wriggle, thrash and arch up – the sight of this seething mass can make hungry predators think twice. Spitfires are a group of sawfly larvae that take personal safety a step further. They extract potent oils from the plants upon which they feed, store them in sacs attached to their foregut and regurgitate this burning fluid if threatened.

Goodie or baddie?

Sawflies are among the most destructive members of the Hymenoptera, due to the plant-feeding behaviour of their larvae. They attack a wide variety of native, ornamental and economically important trees and shrubs, and a few species make pests of themselves in agricultural and forestry industries.

Fascinating facts

- It can be difficult to tell the difference between caterpillar-like sawfly larvae and the true caterpillars of butterflies and moths. Both have cylindrical bodies and well-developed head capsules and may have suction cups (prolegs) along their abdomens for added grip while foraging on plants. In general, the larvae of sawflies have a bigger head capsule and their bodies gradually taper down towards the end of the abdomen. The larvae of butterflies and moths have smaller head capsules and their bodies are more or less cylindrical (see pp. 68–70 for illustrations of these different larvae).

- When is a slug not a slug? Cherry and pear slugs are the slimy-looking larvae of a particular sawfly accidentally introduced to Australia. Their slippery bodies lack any discernible appendages such as legs, eyes or antennae, making them very slug-like in appearance. As their name suggests, these guys are minor pests on several types of orchard trees.

Learn more

In this book:

Leafblister sawflies, pp. 148–149
Spitfires, p. 127
Steelblue sawflies, p. 162

In other books:

Grissell E (2010) *Bees, Wasps and Ants: The Indispensable Role of Hymenoptera in Gardens.* Timber Press, Oregon. Detailed information on the role of bees, wasps, ants and sawflies in our environment. Features great colour photographs and is an excellent book for gardeners. ●

An explanation of the above book ranking system can be found in Further Reading.

Online, type in:

Spitfire grubs – information and pictures on sawfly larvae that regurgitate toxic substances to ward off predators.

A female raspberry sawfly (*Philomastix macleaii*) stands guard over her eggs. Source: A. Hiller.

This eucalypt-feeding sawfly caterpillar (*Lophyrotoma* sp.) rears up its black needle-like tail when disturbed.

Beetles – Order Coleoptera

Summary – Beetles

Order Coleoptera, meaning 'sheath wing'

359 891 species worldwide, over 28 200 species in Australia

Key features
- Heavily armoured bodies
- Often round or oval in shape
- Fore wings form a shell-like cover that sits tight and flat over the hind wings and abdomen; hind wings are not visible until they fly
- Fore wings meet in a straight line down their backs
- No cerci (finger-like appendages) at the end of the abdomen

Cowboy beetle (*Chondropyga dorsalis*).

Don't confuse beetles with:

Cockroaches

Order Blattodea, see p. 207

Things in common

- Some cockroaches are similar to beetles in their size and shape

Differences

- Cockroach wings look leathery (as opposed to the hard, often shiny, wing covers of beetles) and have a dense network of veins not seen in beetles
- Cockroaches have a pair of finger-like structures called cerci at the end of their abdomen, which are absent in beetles

Balta cockroach (*Balta* sp.).

Earwigs

Order Dermaptera, see p. 219

Things in common

- Fore wings form a shell-like cover that folds tight and flat over the hind wings

European earwig (*Forficula auricularia*). Source: T. Daley.

and meet in a straight line down the back; hind wings are not visible

Differences

- In earwigs the elytra (wing covers) are generally much shorter than those of beetles (although there are a few groups of beetles that have shortened wings)
- Earwigs possess a large pair of pincer-like forceps at the end of their abdomen not usually seen in beetles

Stink bugs, water bugs, plant bugs and their allies

Order Hemiptera, Suborder Heteroptera, see p. 292

Things in common
- Some true bugs are similar to beetles in their size and shape
- Fore wings form a shell-like cover that sits tight and flat over the hind wings and abdomen; hind wings are not visible
- No cerci at the end of the abdomen

Zebra gum tree shield bug (*Poecilometis histricus*).

Differences

- The fore wings of true bugs overlap each other forming a cross, or no line is visible at all, whereas in beetles the fore wings meet in a straight line (see Question 6 in identification key, p. 35)
- True bugs have a needle-like mouth whereas beetles have chewing jaws, but this can be difficult to see

Description

Beetles comprise one of the largest and most diverse Orders of insects in the world. They can be bigger than the palm of a human hand or so tiny they could pass through the eye of a needle with room to spare. They come in a dazzling array of colours and are among the most vibrant animals on our planet. Some extract their bright pigments from the plants they eat, others are clothed in velvety hairs or shimmering powders. Jewel and Christmas beetles reflect light like a disco ball, gleaming metallic shades of gold, silver, red and green. But it is the unique modification to their wings that sets beetles apart from all other insects and is responsible for their vast success.

Instead of two pairs of flapping wings like a butterfly or dragonfly, beetles have taken their fore wings and transformed them into hardened shell-like covers known as elytra, which fold neatly over their delicate hind wings and abdomen. This customised body armour allows beetles to venture into places, such as burrowing through gravel and mud, diving underwater or tunnelling through timber, usually off-limits to winged insects.

At the same time, their flexible hind wings mean that beetles can still fly.

Beetles usually have large compound eyes and are armed with strong, scissor-like jaws known as mandibles. These can be used for slicing and crushing food, boring holes into wood or plants for depositing their eggs, or for defence against enemies. Their antennae come in a variety of shapes. They may be long and thin, linked together like the beads in a necklace, spread out like the teeth of a comb, or gradually thickened into a club.

Beetle legs are generally used for walking or running, but some species have legs modified for swimming, digging and occasionally jumping. The tips of the legs are adorned with a pair of claws, which can be used for grasping and climbing.

Diet and habitat

It is hard to find an environment where beetles are **not** present. Their hardened bodies protect them from harsh conditions and minimise dehydration, so beetles can inhabit the hottest deserts, the steamiest rainforests and the most frigid caves and mountain ranges. Some beetles live underwater as both adults and larvae, others burrow through fresh dung, rotten carcasses and decomposing vegetation. In any tree you can find beetles in the foliage, fruit, flowers, seeds, branches, trunk and roots. Some beetles exploit other animals, loitering around the nests of birds or lurking in the tunnels of ant and termite colonies. They even infest the flour and pasta in our pantries.

Many beetles feed on plants and some are quite fussy about the type or particular part of the plant they will eat. Leaf beetles, Christmas beetles and many weevils happily devour leaves, often completely defoliating the tree in the process. Wood-boring beetles such as longicorn and jewel beetles have larvae that burrow through the trunks and branches of trees.

Some beetles are predators, swiftly devouring any prey that is small or slow enough to capture. Ground beetles favour snails, worms and caterpillars and can even overpower small frogs. Diving beetles glide through the water targeting aquatic insects, tadpoles and small fish. A bubble of air tucked underneath their wing covers serves as a scuba tank, allowing them to breathe underwater for long periods while hunting.

Many beetles are scavengers. Rove and carrion beetles feed on the carcasses of dead animals, and dung beetles – well, you know what those guys eat! Some of our native dung beetles are so obsessed with the freshest possible dung that they go straight to the source, using their strong claws to hitch a ride on the nether regions of wallabies and simply dropping onto the poo as it falls from the animal.

Lifecycle

Winning the affection of a female is a serious business in beetles. Male beetles often have long elaborate antennae to capture the special chemical signals released by females during courtship. Rhinoceros and stag beetles must fight to prove their worth, using large horns on their heads and thorax for battle. A ferocious wrestling match results in one suitor being tossed away or flipped helplessly on to his back, while the winner 'gets the girl'.

The number of eggs laid by the female beetle can range from one or two through to thousands, depending on the species. These eggs may be scattered randomly on the ground, carefully glued to leaves or tucked into elaborate burrows provisioned with food for the young, depending on the type of beetle.

Beetles undergo complete metamorphosis (a lifecycle similar to that of a butterfly) with an egg, larva, pupa and adult stage. The larvae of beetles are often referred to as grubs and vary greatly in appearance (see Chapter 3, p. 69). Ladybird beetle larvae are predators, using their slender bodies and long legs to actively chase prey. Scarab larvae (including rhinoceros, cane and Christmas beetles) are found in soil and rotting wood and have a distinctive C-shaped body and long legs, and are clothed in fine hairs. The grubs of weevils have no legs at all. Regardless of their appearance, all beetle larvae have strong mandibles, a voracious appetite and a flexible skin that is shed at each moult to accommodate their growing bodies.

The larval stage is the longest stage in a beetle's life. Larvae that feed on a limited food source, such as the foliage of seasonal plants or the carcass of a dead animal, rush through the larval stage before their food runs out. However, grubs that burrow through timber have all the time in the world and often spend several years growing as larvae.

The larva usually transforms into a pupa in a protected place, such as inside the tree it feeds on, or hidden underground, before emerging as an adult.

Defence
Beetles make up the diet of many animals such as birds, spiders, frogs, lizards and predatory insects. Camouflage is a common means by which they avoid such enemies, and many beetles have colours that allow them to blend into their surroundings. Some pie-dish beetles have taken it a step further – they grow organisms such as lichen on their bodies to help them blend into the forests where they live. When disturbed, some beetles (e.g. weevils) feign death or drop suddenly to the ground to confuse would-be predators.

Some species of beetles take a more aggressive approach to deterring enemies. Some use sharp mandibles and spiky legs to fight their way out of trouble. Others are poisonous or distasteful and communicate this to animals through their brightly coloured bodies. Colour combinations of black with red, orange or yellow are a universal sign of being dangerous or bad to eat. Many predators are reluctant to eat insects that are capable of inflicting a painful sting – some longicorn beetles take advantage of this by copying the bright colours and clear wings of wasps.

The prize for the most effective (and aggressive) way of deterring enemies would have to go to the bombardier beetle. These small ground-dwelling beetles have an arsenal of potent chemicals housed in separate chambers within their bodies. When the beetle is threatened, these compounds are squirted together from the body. As they mix they produce a massive chemical reaction that explodes with an audible pop from the end of their abdomen at over 100°C. Whoever is on the receiving end will be blasted with a boiling cloud of pungent spray, guaranteeing a hasty retreat for the bombardier beetle. And this little guy is not a one-trick pony. This process can be repeated as often as needed – some beetles have been reported to let off 80 'bombs' in as little as four minutes!

Goodie or baddie?
The jury is out on this one. Some species of beetles are serious pests, attacking fruit and vegetable crops, tunnelling through timber

and infesting food products such as flour, dried fruit and spices.

However, many beetles are used as biological control agents in the war against insect pests and weeds. Ladybird beetles and their larvae keep populations of plant-feeding scale insects, mealybugs and aphids under control. A tiny weevil has been introduced from South America to Australia to control salvinia, an invasive aquatic weed that chokes our rivers, lakes and wetlands. Many species of dung beetles have also been released into Australia to deal with the dung produced by cattle and thus keep dung-breeding fly populations in check.

Future Fido?

Some beetles and their larvae make interesting low-maintenance pets (obviously not the bombardier beetle, though!). Mealworms, the larvae of darkling beetles, are available from pet shops and aquariums and are easily reared on a diet of bran and carrots. Rhinoceros beetles are a favourite with kids, due to their large size and quirky horns. The adults generally appear in the warmer months, and when provided with a tank of moist soil and a steady supply of rotten fruit can live for several weeks or months.

Fascinating facts

- If you lined up one representative of every animal and plant species on Earth in a row, around one in five would be a beetle.
- Dung beetles, or scarab beetles as they are sometimes called, are very beneficial insects. Ancient Egyptians worshipped dung beetles as early as 2000 BCE and their likeness can be found decorating the walls of burial chambers, carved on pottery and written in hieroglyphs. The sun god, Ra, was depicted as a dung beetle, rolling the sun across the heavens like a beetle rolling a dung ball. Without dung beetles, our world would be a very smelly place to live in! By gathering up dung and burying it underground for their larvae to consume, dung beetles keep waste levels from cattle in check, reduce food sources for pesky flies and maggots, and keep our soil healthy and well fertilised.
- Certain species of beetles are known as ant-lovers or *myrmecophiles*, as they live within the nests of ants. Some are scavengers, feeding on rubbish and debris within the nest chambers. Others are predators, feeding on the ants themselves or on tiny parasitic mites on the bodies of ants. Some ant-loving beetles secrete sugary substances that ants go crazy for. The ants happily feed the adult beetles and care for their young in exchange for these appealing secretions.
- Have you heard the tragic love story about the beetle and the beer bottle? Females of the giant jewel beetle *Julodimorpha bakewelli* of Western Australia have shiny brown wings covered in little dimples, similar in colour and texture to the bottom of some old-fashioned beer bottles. In roadside areas where littering is common, male jewel beetles come across discarded beer bottles, assume they are enormous super-attractive females and begin to mate frantically with them. So besotted are the males with their artificial girlfriends that the real females are completely ignored and their eggs often go unfertilised as a result. This, combined with the foxes and ants that devour the lovesick males as they swarm over beer bottles, has a detrimental effect on the survival and reproduction of this beetle.

As amusing as it is, this story illustrates the negative impact human actions such as littering can have on the survival of a species. But, thankfully, this story has a happy ending. When beer producers heard of the negative impact their bottle designs were having on these easily confused beetles, they changed the design of the bottle to smooth glass, thus squelching the interest of the amorous males.

- The hardened elytra of beetles are quite tough, especially in some weevils and jewel beetles. So impenetrable are the bodies of these insects that coleopterists (scientists who study beetles) must use a fine jeweller's drill to pierce the bodies so they can pin out specimens for examination.

The mummy-like pupa of a rhinoceros beetle (*Xylotrupes ulysses*), inside which the larva transforms into an adult beetle.

Golden green stag beetles (*Lamprima latreillii*) love to feast on rotting fruits. Source: K. Ebert.

Learn more

In this book:

Cane beetles, p. 87
Christmas beetles pp. 6, 88

The white fluffy-looking larvae of the mealybug ladybird (*Cryptolaemus montrouzieri*) are used in the biological control of many insect pests, including these cottony urbicola scale insects (*Pulvinaria urbicola*). Source: D. Papacek.

INSECT ORDERS 193

Look twice! This wasp-mimicking longicorn beetle (*Hesthesis* sp.) has warning colouration and greatly shortened elytra to show off its transparent hind wings, making it a very convincing wasp mimic. Source: John Tann, Attribution 2.0 Generic.

Cigarette beetles, p. 74
Curl grubs, p. 103
Diving beetles, p. 93
Fiddler beetles, p. 293
Figleaf beetle larvae, p. 148
Flower beetles, p. 134
Leaf beetles, pp. 114, 148
Longicorn beetles, p. 219
Museum beetles, p. 78
Poinciana longicorn beetles, p. 207
Rhinoceros beetles, p. 86
Rice weevils, p. 75
Weevils, p. 115
Whirligig beetles, p. 94

In other books:

Evans AV, Bellamy CL (2000) *An Inordinate Fondness for Beetles*. University of California Press, London.
A fascinating book about the biology of beetles and their role in the environment. Contains excellent photographs of some of the most beautiful and bizarre species from around the world. ●●

Hangay G, Zborowski P (2010) *A Guide to the Beetles of Australia*. CSIRO Publishing, Melbourne.
This book contains excellent general information on the biology of beetles as well as descriptions of the different families of beetles in Australia. Simply by flicking through the pages, you should be able to use the excellent colour photographs to allocate a beetle you have found to a Family. There are photographs of commonly encountered species and information boxes on beetles with interesting behaviours or roles in our environment. ●●
An explanation of the above book ranking system can be found in Further Reading.

Online, type in:

Bombardier beetle – learn more about the explosive chemical arsenal these small beetles use to ward off predators and watch videos of their 'bombs' in action.

Jewel beetle beer bottle – see these easily confused male jewel beetles frantically mate with discarded beer bottles they mistake for giant females.

Booklice – Order Psocoptera

Summary – Booklice

Order Psocoptera, meaning 'gnawed wing'

Synonym: Psocodea (currently valid)

5574 species worldwide, 300 species in Australia

Key features

- Tiny
- Two pairs of membranous wings, held roof-like over abdomen; often with no wings
- Bulging compound eyes
- Antennae thread-like

Booklouse (*Ectopsocus* sp.). Source: T. Daley.

Don't confuse booklice with:

Aphids

Order Hemiptera, Suborder Sternorrhyncha, see p. 278

Things in common

- Tiny
- Two pairs of membranous wings, held roof-like over abdomen; often with no wings

Differences

- Aphids have eyes that are much smaller than those of booklice
- Aphids have two (often black) tubes on the topside of their bodies towards the end of their abdomen, which are absent in booklice

Onion aphid (*Neotoxoptera formosana*). Source: K. Ellingsen.

Lice

Order Phthiraptera, see p. 246

Things in common

- Some lice are similar in shape and size to wingless booklice
- Lice are closely related to booklice

Differences

- Lice are always wingless, whereas some booklice have wings
- Lice have antennae that are much shorter and thicker than those of booklice
- Lice have large claws on each leg, which are absent in booklice

Bird louse (Family Philopteridae). Source: J. Dorey.

Description

Booklice, also referred to as barklice and psocids, are tiny, often drab-coloured insects believed to be closely related to lice. They range in size from less than 1 mm up to 10 mm long. Some bigger ones have beautifully marbled wings, while others are tiny, wingless and almost colourless. They are quite common, but usually avoid our detection due to their tiny size and cryptic appearance.

Booklice have a distinctive head, with long thread-like antennae and bulging compound eyes. Their two pairs of wings are usually transparent and held over their abdomen like a little roof or an upside-down V; however, some species lack wings.

Diet and habitat

Booklice prefer moist environments such as the leaves and stems of trees, on or under bark, in leaf litter and mulch and underneath rocks. Some species aggregate in sheltered places, such as the bark of trees, protected underneath a small canopy of fine silk.

Booklice have strange mouthparts, featuring a little rod-like structure that pushes up against their food, acting as a brace. Their gnawing jaws then go to work, scraping and grinding their food like a mortar and pestle. Despite having the word 'lice' in their name, unlike true lice, they do not feed on blood. They prefer to munch on algae, lichen, fungal spores, plant tissue, fragments of dead insects and insect eggs.

Sometimes booklice venture into our homes and can be found clinging to fence palings, running across desks, papers, books and wallpaper and in stored food products such as flour. In fact, the name 'booklice' comes from the tendency of some species to infest damp books and paper, where they feast on mould.

Lifecycle

Male booklice impress the ladies with their dancing skills, performing a nuptial dance as part of their courtship ritual. Music plays an important role in some species, with males clicking a little organ on their hind coxae (hip joints) in a type of song. Females get in on the act, tapping their abdomen against the ground in response.

Eggs are laid on leaves or bark, with some females covering them with a tiny protective sheet of silk. Booklice undergo incomplete metamorphosis and the hatchling nymph looks like a miniature wingless version of the adult. It moults its skin five or six times on its journey to adulthood.

Defence

Being tiny and completely harmless, booklice fall prey to many predators. Not only are they gobbled up by spiders, pseudoscorpions, insects, small birds, frogs and geckoes, but they are also attacked by parasitic roundworms and fungi.

Booklice defend themselves against predators by avoiding detection. Most species are drab coloured, helping them to go unnoticed by enemies. Others shelter under the bark of trees, or spin silken sheets under which to hide. Some nymphs glue small particles of debris to special hairs on their bodies to help them blend into their surroundings.

Goodie or baddie?

Most booklice live outdoors, where they have no negative impact on humans. However, the species that inhabit our homes can cause minor damage to stored food and paper products.

Fascinating facts

- Despite their tiny size, booklice have impressive dispersal capabilities and can be found at very high altitudes and drifting on air currents above the ocean at great distances from land. In fact, some species have been collected from altitudes as high as 3000 m above sea level.
- The Greek meaning of Psocoptera literally translates to 'gnawed wing'. However, the name isn't actually referring to their wings but to the gnawing and rubbing action of their mouthparts as they feed. So, a more correct translation would be 'winged insects that gnaw'.
- Booklice and entomologists (scientists who study insects) are sworn enemies. These tiny insects invade insect collections, chewing their way through delicate butterfly wings and other precious specimens. Due to their tiny size, they often go unnoticed until they have caused significant damage. Collectors must regularly add naphthalene flakes (mothballs) to insect cases to deter booklice.

Learn more

In this book:

Booklice, pp. 247, 279

In other books:

Rees D (2007) *Insects of Stored Grain: A Pocket Reference.* CSIRO Publishing, Melbourne.
Information and pictures on the lifecycle, distribution and biology of some economically important species of booklice which infest stored products such as flour and grains. ●

Robinson WH (2005) *Urban Insects and Arachnids: A Handbook of Urban Entomology.* Cambridge University Press, Cambridge.
A detailed guide to cosmopolitan species of insects and arachnids, including booklice, that invade our homes and backyards. Contains black and white drawings of commonly encountered species. ●●

An explanation of the above book ranking system can be found in Further Reading.

Online, type in:

Barklice webbing – interesting photographs of a species of booklouse that form large sheets of silken webbing on the trunks of trees.

This beautifully patterned booklouse (*Myopsocus* sp.) feeds on lichen and algae growing on wooden surfaces such as fence palings, wooden handrails and the bark of trees. At 6 mm long, it is one of the larger species of booklice in Australia. Source: T. Daley.

Butterflies and moths – Order Lepidoptera

Summary – Butterflies and moths

Order Lepidoptera, meaning 'scaly wing'

Synonyms: Frenatae, Macrolepidoptera, Microlepidoptera, Jugatae

156 793 species worldwide, over 21 000 species in Australia

Key features

- Wings and bodies covered in furry scales
- Two pairs of large wings that are usually solid in colour and not transparent (although there are some exceptions)
- Large compound eyes
- Antennae clubbed (butterflies) or long and thin or feathery (moths)

Butterflies*

Key features

- Antennae clubbed
- Generally fly during the day
- Often brightly coloured
- Usually rest with their wings out flat or held up over their bodies

Black jezebel (*Delias nigrina*). Source: K. Ebert.

Moths*

Key features

- Antennae long and thin or feathery, not clubbed
- Often nocturnal
- Often not as bright in colour as butterflies
- Rest with their wings out flat or held roof-like over their bodies

Hercules moth (*Coscinocera hercules*). Source: A. Hiller.

*Unlike some other insect Orders, butterflies and moths cannot be simply sorted into their separate Suborders. Their classification is quite complicated and, for this reason, no further rankings (e.g. Suborders) have been provided here.

Don't confuse butterflies and moths with:

Bees and wasps

Order Hymenoptera, see p. 162

Things in common

- Some day-flying moths (such as bee hawk moths, p. 204) mimic the bright colours or transparent wings of bees and wasps

Differences

- Bees and wasps have heads and bodies that may be covered in short hairs, whereas those of moths are covered in fluffy scales
- Bees and wasps have a narrow waist, whereas moths do not

Carder bee (*Afranthidium repetitum*).

Moth lacewings

Order Neuroptera, see p. 241

Things in common

- Moth lacewings are a type of lacewing that have wings clothed in tiny hairs, giving them a moth-like appearance

Differences

- Moth lacewings may have short hairs on their bodies, but these do not give them a furry appearance. Moths have bodies covered in furry scales

While the hairy body and wings from this recently emerged moth lacewing (*Ithone* sp.) are still wet and slightly crumpled, you can still see its striking resemblance to a moth. Source: J. Duley.

- Moth lacewings have wings that are hairy, whereas moths have wings covered in flattened scales and therefore look velvety or fluffy rather than hairy
- The wings of moth lacewings have many more veins than the wings of moths

Description

Butterflies and moths comprise a large and easily recognised Order of insects and come in an enormous variety of shapes, sizes and colours. They range from tiny blue butterflies that could fit on a postage stamp with plenty of room to spare, through to massive swift moths and emperor moths whose wings stretch over 25 cm. Some moths may have a wingspan of just 3 mm, meaning two such moths (and they are fully grown adults!) could fit side by side on a single grain of rice.

The one thing all butterflies and moths have in common is tiny flattened scales on their wings and bodies. It is these scales that give them amazing colours and patterns; they easily rub off on your fingers like a fine powder when touched.

Butterflies and moths have four large flapping wings that are sometimes linked together with tiny hooks to help them fly.

They have large rounded compound eyes and a pair of long antennae. Many have a long coiled tube for drinking nectar, called a proboscis. Up close, this structure looks and works just like a party blower (minus the noise)! It is rolled up during flight and is unravelled for feeding.

The legs are long and slender to help them balance on flowers and fruit as they feed. Their feet or *tarsi* are covered in special receptors that help them taste, touch and smell.

Diet and habitat

Moths occupy a much wider range of habitats than butterflies and far outnumber them in terms of species. In fact, butterflies make up only around 2% of Australian Lepidoptera, and most of them are confined to tropical environments.

Butterflies and moths prefer a liquid-only diet and use their proboscis to slurp nectar from flowers, sap from plants, juices from rotten fruit and water from puddles. A few primitive species of moths emerge from their cocoons without functioning mouthparts, instead using the stored fat in their enormous abdomens to sustain them long enough to reproduce and lay eggs.

The larvae of butterflies and moths feed on entirely different foods from their nectar-loving parents. Caterpillars use their strong mandibles to feast on the leaves, twigs, roots, flowers, fruits or seeds of plants. Some are conspicuous feeders, clustering together in large groups on foliage to feed. Others are more secretive, hiding away in silken shelters, tunnelling their way beneath the surface of bark or leaves, or feeding deep within the trunks of trees.

Lifecycle

Butterflies use their beautifully coloured wings in courtship to help them recognise and capture the attention of an appropriate partner for mating. Often the males are the most beautiful, using their bright patterns and showy aerobatics to win the affection of females. Up close, the male releases a cocktail of special chemicals known as pheromones, which are used by the females to confirm his identity as a potential mate (different species have different pheromones).

Moths are mostly nocturnal and visual displays would not be effective in the dark. Instead, the females release strong pheromones that drift along air currents. Males often have elaborately feathered antennae, increasing the surface area of these sensory appendages so they can more easily pick up these smells. Emperor gum moths can detect pheromones from a female up to a kilometre away.

Butterflies and moths undergo complete metamorphosis, with their lifecycle split into four distinct stages – egg, caterpillar (larva), cocoon or chrysalis (pupa), and adult. This cycle may take anywhere from a few weeks to several years, depending on the species.

Despite being miserable parents which offer no care for their young, many moths and butterflies go to great lengths to ensure they lay their eggs on a plant that will give their tiny larvae the best possible start in life. Adult females track their usual host plants from a distance, locking onto special chemical smells released by the plant to pinpoint its location. Once on the plant they scratch at the surface of a leaf, using special taste receptors located in their feet to sample the chemicals released to evaluate its suitability as a host for their young. The eggs are then glued securely to the leaf, either singly or in batches.

Caterpillars have an armoured head capsule, strong chewing mandibles and thick cylindrical bodies (see pp. 69–70).

They eat voraciously, ensuring they have enough food stored in their bodies before they wriggle out of their final larval skin, revealing a pupa. Moth larvae often weave a soft protective bag of silk around themselves before entering the pupal stage – this is called a cocoon. The pupae of butterflies are generally left naked (referred to as a chrysalis), often suspended from the plant with a strong girdle or pad of silk. They look inactive, but inside great changes are taking place. Their bodies break down the tissues, reorganise them and regrow into the adult form.

Defence

With their fat juicy bodies, butterfly and moth larvae are very appealing to predators such as birds, and for this reason have developed many ways to avoid being eaten. Caterpillars are masters of mimicry, with some using false eyes on their bodies to impersonate larger animals such as reptiles or birds. Others are poo mimics, using their shiny brown and white markings to resemble an unappetising bird-dropping.

Keeping your location hidden is another tactic to avoid the attention of predators. Camouflage is common, with many species resembling leaves, sticks, bark and lichen. Leaving droppings (known as frass) may help a predator or parasite to pinpoint the location of a caterpillar. For this reason, some caterpillars use explosive defecation, the process by which their droppings are flung away from their bodies with surprising force. Some cabbage white caterpillars grasp frass in their strong jaws and fling it away, while the larvae of skippers catapult their faeces from the ends of their abdomens, up to 1 m from their bodies.

Several caterpillars take a more aggressive approach to warding off enemies.

Many have itchy or stinging hairs that produce pain, rashes or allergic reactions when touched. Some brightly coloured caterpillars are poisonous, extracting chemicals from the leaves they feed on and converting them into powerful toxins, which they store inside their bodies, even as adults. Swallowtail caterpillars equip themselves with a pair of nasty-smelling, inflatable red 'devil horns' known as osmeteria which they keep tucked in a special pouch behind their heads. The slightest touch from an inquisitive bird triggers their release and generally leads to the enemy's hasty retreat.

Butterflies and moths are usually large and highly visible, making them an easy target for predators. Some butterflies, such as skippers, have an erratic way of flying and can often outmanoeuvre a hungry bird. Many moths are experts in camouflage, disappearing into foliage, bark and leaf litter when at rest.

Startle displays are also common. Many butterflies are well camouflaged when their wings are shut but if disturbed flash them open, revealing dazzling colours and patterns that can frighten or disorient would-be predators. Others have large false eye-spots on their wings. When peering into the bushes to forage, a bird will be confronted by what it assumes is the face of a much larger animal, causing it to flee.

Goodie or baddie?

This is a tricky one. Fruit-piercing moths can be a pest in orchards, plunging their proboscis into the skin of ripe fruit and sucking out the juices. However, most butterflies and moths stick to nectar and are such a joy to see in our gardens that we rarely care if the odd one causes a few bruises in our fruit. They can also play a significant role in pollinating our plants.

Caterpillars are another story. Many have developed a taste for our vegetable gardens, such as those of the cabbage white butterfly that tear their way through cabbages, broccoli and cauliflower. And anyone who has strayed too close to a hairy caterpillar will recall the itchy and often painful rashes or allergic reactions that follow.

But in their defence, if it wasn't for the tiny caterpillars of the cactoblastis moth, millions of hectares of land in Queensland and New South Wales would still be overrun with prickly pear. The release of this moth from the Americas in the late 1920s is still regarded as the world's biggest success story in the biological control of a weed.

Future Fido?

Watching a caterpillar grow and develop into a beautiful butterfly is an amazing experience. Many, such as orchard swallowtails, are easily found in the garden and reared in captivity, allowing you to witness this transformation firsthand. However, do your research first. Some species accept only a single kind of plant – without it, your caterpillars will perish.

Fascinating facts

- Moths talking to bats?! Moths have hearing organs on the underside of their bodies and some moths use these to listen for the ultrasonic calls bats emit to locate their prey while hunting. Once alerted to a bat's presence, the moth can take evasive action such as dropping suddenly to the ground or adjusting its flight to outmanoeuvre these aerial enemies. Some tiger moths talk back, clicking a special structure on their thorax to advise the bats that they are distasteful to eat.
- Some butterfly caterpillars are carnivorous and happily devour scale insects, ant larvae and the eggs of other butterflies.
- Moths from the Genus *Calyptra*, found in tropical Asia, are known to drink the blood of cattle, elephants and occasionally humans! The males of this so-called vampire moth use their proboscis to pierce the skin of their victim and suck up a meal of blood. It is believed they extract salt from the blood and offer it as a romantic gift to female moths, which are often lacking in sodium as a result of egg production. Other species of moths drink fluid

What you can do – egg hunt

Butterflies may flutter around a particular plant in your garden, swooping down to land on individual leaves briefly, before taking to the air again. This behaviour is known as *alighting* and preludes the laying of eggs. If you search the leaves where a butterfly has alighted, you may find their tiny eggs. The size, shape and colour of these eggs vary between different species of butterflies. As a rough guide, they are usually spherical or oval, around the size of a mustard seed (if you have more of a sweet tooth – about as big as hundreds and thousands sprinkle) and are often pale.

- discharged from the eyes of animals (e.g. cattle), presumably for the same purpose.
- Moths leave nothing to chance. Female goat moths lay more than 18 000 eggs, ensuring that at least a few survive the perilous journey to adulthood.
- There are around 17 461 species of butterflies in the world. This pales in comparison to the number of moth species, which is close to 140 000.
- How many moths? The exact number of species in Australia is very difficult to estimate as several thousand newly discovered moths currently sit in insect collections around Australia, waiting for someone to undertake the enormous task of describing them and assigning each a unique scientific name.

Learn more

In this book:

Moths
Case moths, pp. 155–157
Citrus leafminers, p. 149
Clothes moths, p. 85
Corn earworms and native budworms, p. 121
Granny's cloak moths, p. 82
Hawk moth pupae, p. 104
Indian meal moths, p. 76
Lawn armyworms, p. 128
Pasture webworms, p. 243
Processionary caterpillars, pp. 138–139
Scribbly gum moths, pp. 149–150
Silk moths, p. 37

Butterflies
Blue tigers, p. 141
Cabbage whites, p. 120
Cairns birdwings, p. 15
Caper whites, p. 140

Common crows, pp. 157–158
Swallowtails, pp. 118–119
Tailed emperor, p. 23
Wanderers, p. 37

In other books:

Braby MF (2016) *The Complete Field Guide to Butterflies of Australia.* 2nd edn. CSIRO Publishing, Melbourne.
A full-colour field guide to the different species of Australian butterflies. Life-sized photographs and distribution maps make this guide very user-friendly. ●●

Clyne D (2011) *Attracting Butterflies to Your Garden.* New Holland, Sydney.
A beautifully photographed book containing practical tips for attracting butterflies to your garden. ●

Clyne D (2009) *The Secret Life of Caterpillars.* New Holland, Sydney.
A great book for kids and adults alike. ●

Common IFB (1990) *Moths of Australia.* Melbourne University Press, Melbourne.
A detailed guide to the different families of moths in Australia, including diagrams of key morphological features and lots of black and white and colour photographs of various species. ●●●

Field RP (2013) *Butterflies: Identification and Life History. Museum Victoria Field Guide.* Museum Victoria Publishing, Melbourne.
This field guide specifically covers the 128 species of butterflies found in Victoria. However, as many of these butterflies are also found in other areas of Australia it is worth a read even if you don't live in the region. As well as great information on the distribution, biology and larval food plants of each species, the book includes detailed

colour photographs of adult butterflies, larvae and pupae and (for the first time that I know of in a field guide) spectacular close-ups of the butterfly egg. ●●

Orr A, Kitching R (2010) *The Butterflies of Australia*. Jacana Books, Allen & Unwin, Sydney.

One of my favourites. A very detailed guide to the species of Australian butterflies, containing stunning hand-painted drawings of butterflies, eggs, larvae, pupae and host plants. It also includes handy distribution maps. ●●

Zborowski P and Edwards T (2007) *A Guide to Australian Moths*. CSIRO Publishing, Melbourne.

A full-colour guide to the different families of Australian moths. A detailed chapter on their biology and information boxes on unusual or commonly encountered species throughout the book make it very informative. ●●

An explanation of the above book ranking system can be found in Further Reading.

Online, type in:

Interesting caterpillars – type this into an image search, then sit back and marvel at photographs of some of the world's most beautiful and bizarre caterpillars.

Madagascan sunset moth – this butterfly-mimicking day-flying moth is regarded as one of the world's most beautiful insects.

Monarch butterfly migration – see amazing photographs and footage of huge clouds of these colourful butterflies

With their clear wings and colourful markings, bee hawk moths (*Cephonodes kingii*) are often mistaken for bees as they hover around flowers.

Wanderer (or monarch) butterflies (*Danaus plexippus*) use their bright colours to warn predators that they are toxic to eat. Source: K. Hiller.

'Hairy caterpillars' or 'itchy grubs' like this anthelid moth larva (*Anthela varia*) have hairy bodies that can produce pain, rashes or allergic reactions when touched.

To protect itself from enemies, this Cairns birdwing butterfly larva (*Ornithoptera euphorion*) waves strong-smelling osmeteria (inflatable fleshy horns) from behind its head. Source: K. Hiller.

as they migrate south from the northern states of the US, to escape the cold.

Moth camouflage – type this into an image search and try to spot the moths.

Owl butterfly – do an image search for amazing pictures of this butterfly. Huge circular spots on its wings mimic the giant golden eyes of an owl.

Richmond birdwing butterfly – learn more about the plight of this vulnerable butterfly species and the efforts being taken by local groups to restore its numbers in the wild.

Silkworm/silk production – the use of silkworms to produce silk dates back thousands of years. Learn more about the rearing of silkworms (the caterpillars of a type of moth) and how their silken cocoons are processed into textiles.

Croker's frother moth (*Amerila crokeri*) gets its name from the frothy orange bubbles it sprays from special glands on its thorax when disturbed. Source: A. Hiller.

Cockroaches – Order Blattodea

Summary – Cockroaches

Order Blattodea, meaning 'insects that shun the light'

Synonyms: Dictyoptera, Blattaria

4565 species worldwide, 534 species in Australia

Key features
- Flattened, oval-shaped bodies, often brown
- Long spiky legs
- Leathery fore wings loosely cover membranous hind wings; wings sometimes absent
- Large compound eyes
- Antennae thread-like
- Two cerci (finger-like appendages) at the end of the abdomen

Ellipsidion cockroach (*Ellipsidion reticulatum*).

Don't confuse cockroaches with:

Beetles

Order Coleoptera, see p. 186

Things in common
- Flattened, oval-shaped bodies, often brown
- Some beetles have thread-like antennae

Differences
- The fore wings of cockroaches have a very distinct network of tiny veins which beetles lack
- Beetles have no cerci at the end of the abdomen, whereas cockroaches have two cerci

Poinciana longicorn beetle (*Agrianome spinicollis*).

Stink bugs, water bugs, plant bugs and their allies

Order Hemiptera, Suborder Heteroptera, see p. 292

Things in common

- Flattened, oval-shaped bodies, often brown

Differences

- True bugs have wings that form tight-fitting leathery covers over their bodies and these do not reach past the end of their abdomens. Cockroaches' wings loosely cover their abdomens and extend past the ends of their bodies
- True bugs have no cerci at the end of the abdomen and have thicker shorter antennae than cockroaches

Crusader bug (*Mictis profana*).

Description

Cockroaches are small to medium-sized insects (up to 70 mm long) with flattened, oval-shaped bodies that are often brown. They have two pairs of wings, with the first pair forming a stiffened cover known as tegmina, which protect the fan-like hind wings. Many species lack wings, particularly those that burrow underground. The heads of cockroaches are partially covered by their shield-like thorax and feature large kidney-shaped eyes and strong chewing mandibles.

Cockroaches have long spiky legs used for running and sometimes digging. Some species have a remarkable suction cup between the claws on their feet that is used for moving across slippery surfaces such as glass and tiles. When special muscles in the feet are flexed, these pads automatically inflate like miniature air-bags, spreading out and adhering to the surface. When the muscle is relaxed, the suction cups deflate and tuck neatly back between the claws of the feet.

The sensory organs in cockroaches are quite remarkable. Their long thread-like antennae are covered in as many as eight different types of specialised hairs, called setae, which can be used to taste, touch, evaluate the temperature and humidity of their surroundings and smell for both food and potential mates. At the other end of their bodies a pair of small finger-like structures, called cerci, protrude from the tip of the abdomen. These are also covered in tiny sensitive setae to help cockroaches detect their surroundings.

Diet and habitat

Cockroaches are generally found in the bush, either wandering around the foliage of plants by day, or sheltering among bark, logs, leaf litter and rocks. Unfortunately, cockroaches have taken very well to urban environments and have become frequent (and unwelcome) guests in our houses, drains, rubbish bins and compost heaps. Most of these suburban species are nocturnal, sheltering in warm dark places during the day and emerging at night to gorge themselves.

Of all the insects, cockroaches are the **least** fussy of eaters. In the wild, they are usually vegetarians, munching on leaves, bark and decomposing plant material. But in our food-laden houses they consume any type of foodstuffs, as well as paper, leather, clothes, sewage, book bindings, soap, plastic, polystyrene and – wait for it – the eyelashes, fingernails and dead skin flakes of sleeping humans. Eeek! Cockroaches enjoy a good drink too and happily lap up water from our sinks, dripping taps and toilet bowls.

As you may imagine, a crazy diet such as this can wreak havoc on the digestive system of a cockroach. To make the most of their meals, cockroaches have tiny organisms known as protozoans living in their gut. These secrete enzymes to help break down tough materials such as the cellulose in plant tissue.

Lifecycle

Cockroaches lay their eggs in a little purse-shaped structure, called an ootheca, which holds between 12 and 40 eggs. Oothecae (pl.) may be scattered on the ground, glued to surfaces or carried around within the mother's body (you can sometimes see them protruding from the end of the body).

To hatch, the baby cockroaches (nymphs) must work as a team. They swallow gulps of air, causing their bodies to inflate like tiny balloons. Their rapidly expanding girth prises apart the walls of their ootheca, allowing them to escape into the outside world. The nymphs undergo incomplete metamorphosis, resembling their parents upon hatching then moulting their skin several times to reach adulthood.

Sometimes we come across pure white cockroaches but they are not, as some people assume, albino cockroaches. These ghostly white insects are simply cockroaches which have recently shed their skins. It takes several hours for their exoskeleton to harden and darken to become the usual brown colour.

Female cockroaches can be surprisingly good mothers and show different degrees of parental care, depending on the species. For example, a German cockroach female carries her ootheca protruding from the end of her abdomen until the eggs hatch. She then ushers the tiny nymphs under her wings, protecting them until they are ready to disperse.

Defence

We all know how hard it is to hunt and kill a cockroach. They have a myriad of defensive techniques to avoid being trodden on, sprayed or squashed with a newspaper.

Their legs, antennae and cerci are covered with thousands of tiny sensitive setae, acting as an early warning system. At the smallest hint of danger, messages are sent directly to their incredibly flexible legs and the cockroach is outta there! The reaction time for cockroaches has been measured at an astounding 40 ms (that's 40 one-thousandths of a second)!

The bodies of many household cockroaches are covered in a slippery substance, allowing them to wriggle into impossibly small cracks and crevices to escape danger. This waxy covering gives large infestations of certain cockroaches their distinctive (and utterly horrible!) odour.

In the wild, cockroaches are masters of camouflage, often mimicking bark and leaves. A few species are brightly coloured and may be toxic or distasteful to predators. Some cockroaches can produce loud hissing noises to scare off would-be attackers, or squirt foul secretions from the ends of their bodies if seized.

Goodie or baddie?
To most people, household cockroaches are among the worst offenders out there. They are considered dirty because they invade our food, carry bacteria on their spiky legs and sometimes trigger allergic reactions in humans when present in large numbers. But bear in mind that with more than 4000 species of cockroaches on our planet, only around 35 species have left their bushland environments to wreak havoc in our homes. Most cockroaches are wonderful; they feed on vegetation and leaf litter in natural habitats and many play an important role in recycling nutrients into the soil.

Future Fido?
You wouldn't believe it, but some cockroaches make great pets. The giant burrowing cockroach is a native Australian species that feeds on dead gum leaves. They can be kept in an aquarium of dirt or sand, where they happily burrow around.

Fascinating facts
- Cockroaches have very sensitive hearing organs located in their knee joints. Research suggests that these highly efficient 'ears' can detect earthquakes as tiny as 0.07 on the Richter scale.
- Pest cockroaches are so tough they can survive without a head! They mate, lay eggs and walk around minus their heads until they starve to death (which may take several weeks).
- Giant burrowing cockroaches are so large and heavily armoured that they are sometimes mistaken for small freshwater turtles as they are seen crossing roads at certain times of the year.
- Cockroaches are quite gassy! They can break wind as often as every 15 minutes and may continue to 'cut the cheese' for up to 18 hours after they die. In fact, research suggests that around 20% of methane gas production on our planet is the result of insect flatulence.
- The antennae of a male American cockroach may consist of up to 135 individual segments, covered in over 46 200 specialised sensory hairs – that's around 950 hairs per square millimetre.

Learn more

In this book:

American cockroaches, pp. 80, 257
Balta cockroaches, p. 187
Bush cockroaches, p. 110
Cockroach oothecae, p. 159
German cockroaches, p. 79
Wood roaches, p. 105

What you can do – cockroach grooming

Next time you see a cockroach scuttle across your kitchen floor, don't squash it. Instead, capture it and place it into a brown paper bag with a tablespoon of ordinary household flour. Twist shut the top of the bag and shake well, so that the cockroach is covered in a fine dusting of flour. Remove the cockroach from the bag and place it in a jar or aquarium, making sure it has plenty of room and cannot escape. Cockroaches are fanatical about grooming and over the next hour or so you can watch the captured one twist and turn, using its mouth to remove all traces of flour from its body.

In other books:

Copeland M (2003) *Cockroach*. Reaktion Books, London.

As well as interesting, easy-to-read information on their biology, this book showcases cockroaches as muses for songs, graphic art, fiction and photography. ●

Gordon DG (1996) *The Compleat Cockroach: A Comprehensive Guide to the Most Despised (and Least Understood) Creature on Earth*. Ten Speed Press, California.

Interesting facts and easy-to-read information on cockroaches from around the world. ●

Rentz D (2014) *A Guide to the Cockroaches of Australia*. CSIRO Publishing, Melbourne.

A beautifully photographed, full-colour field guide to the different species of Australian cockroaches. While some of the diagnostic features require the aid of a microscope/magnifier (e.g. the number and arrangements of spines on the legs), simply flipping through the detailed colour photographs and accompanying distribution maps should allow you to allocate a cockroach to a Genus, and in many cases a Species. A great book that anyone can use to learn more about cockroaches. ●●

Rentz D (1996) *Grasshopper Country: The Abundant Orthopteroid Insects of Australia*. UNSW Press, Sydney.

Detailed information on the different families of cockroaches in Australia, including identification keys which can be used (with the aid of a microscope/magnifier) to identify various Subfamilies and Tribes. Colour plates illustrate some commonly encountered Australian species. ●●

An explanation of the above book ranking system can be found in Further Reading.

Online, type in:

Cockroach decapitation – learn how certain cockroaches can survive without a head.

With a body length of up to 75 mm, giant burrowing cockroaches (*Macropanesthia rhinoceros*) are one of the largest species in the world. Source: D. Papacek.

This Sloane's northern wingless cockroach (*Cosmozosteria sloanei*) shelters under bark and logs when disturbed.

Cockroach home remedies – some interesting natural methods for controlling pest species.

Giant burrowing cockroach – these enormous Australian cockroaches make great pets and are one of the largest species in the world.

Dragonflies and damselflies – Order Odonata

Summary – Dragonflies and damselflies

Order Odonata, meaning 'toothed jaws'

5680 species worldwide, 324 species in Australia

Key features

- Elongate slender bodies
- Legs all equal in size
- Two pairs of similarly sized, blade-like membranous wings, with networks of tiny black veins
- Wings held up vertically above the body, or out flat when at rest
- Extremely large bulbous compound eyes that take up most of the head
- Antennae bristle-like and hard to see

Dragonflies, Suborder Anisoptera

Synonyms: Epiprocta or Epiproctophora

Female painted grasshawk dragonfly (*Neurothemis stigmatizans*).

Key features

- Bodies usually large and robust
- Compound eyes cover almost the entire surface of the head
- Hind wings are slightly wider than fore wings
- Usually rest with their wings out flat

Damselflies, Suborder Zygoptera

Key features

- Bodies usually slender and delicate
- Compound eyes are smaller than those of dragonflies and are perched on the outer edges of the head
- All four wings equal in size
- Often rest with their wings held up over their bodies

Bronze needle damselfly (*Synlestes weyersii*).

Don't confuse dragonflies and damselflies with:

Lacewings

Order Neuroptera, see p. 241

Things in common

- Elongate slender bodies
- Two pairs of similarly sized, blade-like membranous wings, with networks of tiny black veins

- Large bulbous compound eyes

Differences

- Lacewings have antennae that are long and thin and easy to see, whereas dragonflies and damselflies have tiny bristle-like antennae that are difficult to see

Blue eyes lacewing (*Nymphes myrmeleonides*).

Description

Dragonflies and damselflies are large flying insects with long, slender, often brightly coloured bodies. They usually range from 3–9 cm in length, but there are a few whoppers out there that can reach up to 15 cm. They have four transparent wings, criss-crossed with a network of tiny black veins. Their thorax is large and rectangular, housing strong muscles used for flight. Their slender spiky legs are angled forwards, helping them to grasp prey in flight and cling to twigs and other perches while at rest. A pair of strong toothed jaws (mandibles) and two short bristle-like antennae may be visible on the head.

Dragonflies and damselflies have extraordinary eyesight thanks to a pair of enormous, multi-faceted compound eyes on their large, highly mobile heads. Different parts of the eye are allocated different tasks, with the front section overseeing their flight while the top of the eye keeps a lookout for enemies, prey and potential mates. Three simple eyes (ocelli) on the forehead monitor the position of the horizon, much like the altitude indicator in a plane, and feed this data directly to the muscles in the wings so flight adjustments can be made.

Dragonflies and damselflies begin life as wingless, aquatic juveniles that differ greatly in appearance from the adults. These larvae are shorter and thicker than their parents and are usually brown or green to help them blend into their surroundings. Their eyes are much smaller than those of adults and their strong legs are used for walking, climbing and burrowing. A series of gills allows these insects to breathe underwater – in damselflies they appear as three delicate leaf-like structures at the tip of the abdomen, but in dragonflies they are hidden within the hind gut.

Diet and habitat

Dragonflies and damselflies are most active in the warmer months and are usually found around aquatic habitats such as ponds, lakes and streams where they come together to mate and lay their eggs. They are often seen perched motionless on rocks and vegetation, darting out occasionally to hunt, mate or chase away intruders (they are highly territorial).

Dragonflies and damselflies are the ultimate aerial predator. They use their superior eyesight to locate prey, grasp it with their long legs while in flight then devour it with their strong jaws. If you have ever seen these insects swooping repeatedly through the air, they are most likely gorging themselves on clouds of mosquitoes and midges.

The larvae are also fearsome hunters, devouring a wide array of aquatic animals such as snails, flatworms, aquatic insects, tadpoles and small fish. They are ambush predators, camouflaging among aquatic plants, hiding behind debris or crouching in the silt until their meal gets within striking distance. And boy, can they strike!

The lower 'lip' of the larva is modified into an extendable grasping organ, with a pair of special sensory structures with razor-sharp tips, known as palps, located at the end. This organ is kept neatly tucked under the chin when not in use and can shoot out at lightning speed to seize and grasp prey.

Lifecycle

Mating in dragonflies and damselflies is often a brutal affair. The male grasps the female's head or thorax with special claspers at the end of his abdomen, often scratching or puncturing her head or eyes in the process. The female arches the tip of her body up to the base of the male's abdomen, which houses his reproductive organs. If you watch the behaviour of these insects near the water, you can often see mating pairs in this distinctive wheel-like posture.

Dragonflies and damselflies undergo incomplete metamorphosis and incorporate both aquatic and terrestrial habitats into their lifecycle. Some dragonflies and damselflies cut a tiny slit in the stems of aquatic plants and place their eggs inside, sometimes descending a metre below the surface of the water to do so. Others simply scatter their eggs in the water, relying on a layer of sticky jelly on the outside to help them anchor to aquatic vegetation.

The larvae, also referred to as nymphs or naiads, feed voraciously and moult several times before crawling out of the water and shedding their final skin to reveal the winged adult. This cycle from egg to adult can take anywhere from a couple of months to several years.

Defence

Not only do dragonflies and damselflies have to avoid the usual enemies of insects, such as frogs, reptiles, birds and spiders, but many are enthusiastic cannibals and thus they spend a great deal of time trying to avoid each other!

Adults are particularly vulnerable to predators while laying their eggs, resting on vegetation and emerging from the water as adults. They occasionally blunder into sticky spider webs. But out in the open, their amazing aerobatics are unrivalled and they can outmanoeuvre hungry birds by rapidly twisting and turning in the air.

Larvae fall prey to fish, frogs and other aquatic insects. They rely heavily on camouflage, trapping particles of mud and sand in the fine hairs on their bodies to aid concealment.

Goodie or baddie?
Dragonflies and damselflies are a pleasant and colourful addition to your garden. They are harmless to humans, control mosquito populations through the feeding behaviour of both adults and larvae, and can be important indicators of water quality in streams, lakes and ponds.

Future Fido?
With their frisky flight, adult dragonflies and damselflies are not suitable to keep in captivity. However, the larvae can make a very interesting pet to keep in an aquatic display. Make sure you provide a few aquatic plants for oxygen, a stick to climb out of the water to emerge as adults (leave them plenty of space to hang from the stick to harden their wings) and a steady supply of mosquito wrigglers, worms or freshwater shrimp as food. Most importantly, keep only one individual per tank – a larva will happily devour its brothers and sisters!

Fascinating facts
- Dragonflies and damselflies have the biggest and most advanced eyes in the insect world. A single dragonfly eye can contain over 28 000 individual facets to help detect light.
- Dragonfly larvae have a series of valves in the hind gut. They use these to draw water into their anus for respiration. They can also shoot the water out with explosive force, rapidly propelling them away from danger or towards prey.
- Dragonflies and damselflies easily become confused. The shiny surface of concrete, bitumen or a freshly washed car resembles the surface of water and can cause females to swoop down and try to lay their eggs.
- Some dragonflies have been clocked flying at over 35 km/h.
- As mentioned earlier, dragonflies and damselflies undergo incomplete metamorphosis and their young may be referred to as larvae, nymphs or naiads. While all terms are considered correct, the word 'larvae' has been used in this book rather than 'nymphs' as this term is used in most of the field guides and books recommended here as further reading.

What you can do – observing dragonflies

Dragonflies are very territorial. Head to your local creek or dam and watch the males vigorously defend their territory by chasing away rival males. Watch the water for females dipping the tips of their abdomens beneath its surface to deposit their eggs.

Learn more

In this book:

Common bluetail damselflies, p. 242
Dragonfly and damselfly larvae, p. 99
Variable tigertail dragonflies, p. 37

In other books:

Brooks S (2002) *Dragonflies*. Natural History Museum, London.
An easy-to-read, beautifully photographed book on the biology of dragonflies (and damselflies) around the world. ●

New TR (1991) *Insects as Predators*. NSW University Press/Australian Institute of Biology, Sydney.
Detailed information on how dragonflies and damselflies select, capture and consume their prey. ●●●

Silsby J (2001) *Dragonflies of the World*. CSIRO Publishing, Melbourne.
A beautifully photographed and very readable book on the biology of dragonflies (and damselflies) around the world, including a detailed breakdown of the different Families. ●●

Theischinger G, Hawking J (2006) *The Complete Field Guide to Dragonflies of Australia*. CSIRO Publishing, Melbourne.
A full-colour field guide to the dragonflies and damselflies of Australia, including distribution maps and diagrams of the key features of each species. A highly detailed illustrated identification key can be used (preferably with the aid of a microscope/magnifier) to sort adults and larvae into various Families, Genera and Species. ●●

Watson JAL, Theischinger G, Abbey HM (1991) *The Australian Dragonflies: A*

A male jewel flutterer dragonfly (*Rhyothemis resplendens*) surveys his territory.

Guide to the Identification, Distributions and Habitats of Australian Odonata. CSIRO Publishing, Melbourne.
A guide to the adult dragonflies and damselflies of Australia, including black and white diagrams, some colour photographs and a detailed illustrated identification key that can be used (preferably with the aid of a microscope/magnifier) to sort adults into various Families and Genera. ●●

An explanation of the above book ranking system can be found in Further Reading.

Online, type in:

Dragonfly eyes – type this into an images search and you will have dozens of amazing close-up photographs of dragonfly and damselfly eyes.

Dragonfly larva hunts newt – amazing footage of the hunting skills of a juvenile dragonfly.

Meganeura – this prehistoric dragonfly (now extinct) prowled the skies around 300 million years ago and had a wingspan similar to that of a seagull.

The large 'chin' on this unicorn darner dragonfly larva (*Austroaeschna unicornis*) extends into a grasping organ, complete with razor-sharp tips for impaling prey. Source: Gooderham and Tsyrlin (2002).

Fiery skimmer dragonflies (*Orthetrum villosovittatum*) mate in a wheel-like posture. The red male grasps the back of the female's head as she arches her body up to the base of his abdomen, where sperm is stored in a special sac. Source: K. Hiller.

You can find the shed skins of dragonflies clinging to the vegetation near water.

Earwigs – Order Dermaptera

Summary – Earwigs

Order Dermaptera, meaning 'skin wing'

1967 species worldwide, 91 species in Australia

Key features

- Elongate flattened bodies
- Legs short, stout and all equal in size
- Fore wings form a short shell-like cover that sits tight and flat over the hind wings and abdomen, wings sometimes absent
- Fore wings meet in a straight line and extend only a short way down the abdomen; hind wings not visible
- Large pair of pincer-like forceps at the end of the abdomen
- Antennae long and thin

Common brown earwig (*Labidura truncata*).

Don't confuse earwigs with:

Beetles

Order Coleoptera, see p. 186

Things in common

- Hardened shell-like covers that fold tight and flat over their hind wings, meeting in a straight line down the back

Differences

- In earwigs, the shell-like covers are greatly shortened, extending only a little way down the back
- Beetles usually do not have pincer-like forceps at the end of the abdomen

Longicorn beetle (*Amphirhoe decora*).

Description

Earwigs are small to medium-sized insects with flattened elongate bodies that reach up to 50 mm. They range in colour from pale brown through to black. Their distinguishing feature is a pair of strong pincer-like forceps at the end of their flexible abdomens. These forceps can assist in a variety of tasks, such as capturing and carrying prey, defence against predators, and courtship displays. They also help neatly tuck away the earwig's wings when not in use.

Earwigs have strong chewing jaws (mandibles) and short equal-sized legs used for running and walking. The wings, when present, are quite remarkable. The hind wings are elaborately folded like a fan first then tucked lengthwise underneath a pair of hardened shell-like covers know as elytra (referred to as *tegmina* in some books) formed by the fore wings.

Diet and habitat

Shunning the light of day, earwigs like to shelter in dark, moist, tight spaces. They can be found tucked underneath the bark of trees, in rotting logs and hiding under leaf litter and other debris on the ground. Humans mostly encounter them at night, attracted into our houses by bright fluorescent lights.

Earwigs have a very relaxed approach to eating. Their diet consists of a wide array of live and dead plant and animal material. Some earwigs are formidable hunters, catching soft-bodied insects such as caterpillars with their strong forceps. Others munch on leaves, flowers, fruit and decomposing plant material.

Lifecycle

Male earwigs can usually be distinguished from the females by their larger and more curved forceps. After mating, the female digs a shallow burrow underneath soil or leaf litter and lays her eggs in a batch at the bottom. The number of eggs per batch depends on the species of earwig and can range anywhere from 20 to 80.

The female cares for both the eggs and newly hatched nymphs, which is quite uncommon in the 'drop your eggs and go' world of insects. After they are large enough to fend for themselves, the nymphs must leave the burrow. Any 'teenagers' that lag behind expecting free food and board are promptly eaten by their mum. No 'Mother of the Year' awards for this critter!

Earwigs undergo incomplete metamorphosis, moulting (and eating!) their old skin several times to become adults.

Defence

The female earwig takes the job of protecting her eggs and young very seriously. She closely guards her burrow, attacking any animal that approaches her nest. She gets so caught up in her vigorous ejecting duties that it is not uncommon for a courting male to get tossed out accidentally.

The large forceps can be used in offence and defence – a well-aimed pinch is often sufficient enough to discourage predators, which include bats, birds and spiders.

Goodie or baddie?

European earwigs (an introduced species) can be a serious pest in our gardens. They tear through a wide array of leafy green vegetables, munch through fruit and attack the petals and buds of flowers such as roses. They can also attack our homes, chewing holes in clothes and carpets.

Fascinating facts

- The name 'earwig' is most likely derived from 'ear wing'. When unfolded, their fan-like hind wings resemble a human ear.
- Earwigs have developed a bad reputation, mainly from the myth that they love nothing more than to burrow into the ears of unsuspecting humans.

Thankfully this is not true, with most earwigs preferring to stretch their legs on the forest floor than cram their way into our ear holes.
- A mother earwig's duties include carefully licking her batches of eggs. This helps to remove fungal spores from their surface, preventing them from going mouldy before they can hatch.
- Males of several earwig families have not one but two penises! Further research has revealed that they are almost always 'right-handed', that is, they mate using the reproductive organs on the right side of their bodies. The left side is a back-up in case of emergencies.

Learn more

In this book:

European earwigs, p. 187

In other books:

See Further Reading

Online, type in:

Earwig urban legends – myth-busting the hilarious tales of ear-burrowing, brain-devouring earwig encounters.

Saint Helena earwig – the world's largest earwig is also one of the most elusive, with no live specimens reportedly sighted since 1967.

A common brown earwig nymph (*Labidura truncata*) showing its curved pincer-like forceps. Source: J. Dorey.

Fleas – Order Siphonaptera

Summary – Fleas

Order Siphonaptera, meaning 'tube without wings'

2048 species worldwide, 90 species in Australia

Key features
- Tiny, usually brown
- Body laterally compressed (tall and skinny) and covered in spines
- Long spiky jumping legs
- Never have wings
- Pair of claws at the end of each leg

Cat flea (*Ctenocephalides felis*). Source: J. Dorey.

Don't confuse fleas with:

Lice

Order Phthiraptera, see p. 246

Things in common

- Tiny, often brown
- Never have wings
- Both lice and fleas are bloodsucking parasites of other animals

Differences

Bird louse (Family Philopteridae). Source: J. Dorey.

- Lice are dorso-ventrally flattened (short and flat) whereas fleas are tall and skinny
- Lice have legs that are shorter, stouter and bear larger claws than fleas

Description

Fleas are tiny brown or black insects that live as parasites on the bodies of other animals. They range in size from 1–10 mm, however, the species we most commonly encounter are usually less than 5 mm long. The bodies of fleas are strongly armoured, laterally compressed (tall and thin) and covered in slicked-back hairs and spines. They keep their short antennae tucked into little grooves on the sides of their head and their sleek bodies are wingless. This streamlined shape helps them push their way through the thick forests of hair, fur or feathers on the body of their host.

Fleas do not have compound eyes, but some species may have simple visual organs known as ocelli. Their mouthparts are highly modified for blood feeding, with sharp serrated blades to pierce the skin of their host and powerful sucking pumps to slurp up blood.

Fleas have long spiky legs that end in a pair of strong claws to help them grip onto the bodies of their host. Their powerful fore legs help them to push their way through feathers or fur, and their enlarged hind legs enable their characteristic jumping behaviour.

Diet and habitat
Adult fleas can be found on the bodies of warm-blooded animals such as mammals and birds. They also spend a great deal of their time off their host and can be found loitering in the nests, burrows and bedding of animals (although due to their tiny size, it is hard to actually see them).

The diet of adult fleas consists solely of the blood of animals. They feed on the bodies of birds, including domestic chickens, pigeons and even sea birds such as petrels. More commonly, they occur on mammals, including mice and rats, bandicoots, cats and dogs, rabbits and wombats. There are even a few species which target aquatic animals, such as penguins and platypus.

The larvae of fleas do not live or feed on other animals. Instead, they hide out in their nests and burrows or in our carpets and pet bedding, where they scavenge on skin flakes, feathers and other organic debris.

Lifecycle
Fleas undergo complete metamorphosis, with an egg, larva, pupa and adult stage. Females lay batches of small, pale, oval-shaped eggs, which may be scattered on the bodies of their host or stuck to surfaces in the nest or burrow. Some species of fleas can produce over 400 of these eggs in their lifetime. The maggot-like larvae use a special tooth on their head to push their way out of the egg and moult their skin several times as they grow, before constructing a flimsy silken cocoon.

If no host animals are present and everything remains still and silent, the newly formed adult simply bides its time, remaining in its cocoon for several months, awaiting the perfect moment to emerge. When the flea senses vibrations, it 'knows' a potential host must be nearby and emerges rapidly from its cocoon. This explains why people often experience large numbers of flea bites when entering a house or room that has been unoccupied for some time.

Once emerged, an adult flea can live for quite a long time. The human flea can live for over 500 days, which is impressive for such a tiny insect.

Defence
As far as we know, fleas in Australia have no natural enemies. However, they do have to contend with being kicked off their host animal as it bites, scratches and grooms its itchy body. For this reason, fleas have well-developed jumping legs – they can simply leap back onto their host if they become dislodged. Fleas can jump up to 20 cm vertically, the equivalent of us leaping over a nine-storey building.

Goodie or baddie?
Without a doubt, fleas can be a major pest in our homes. Human, cat and dog fleas readily bite pets and their owners, causing allergic reactions in some people. More seriously, the bloodsucking behaviour of adult fleas can transmit diseases from animals to humans. While cases of the flea-transmitted bubonic plague are rarely seen in Australia nowadays, we are prone to murine typhus, an infection caused by

bacteria transferred to humans by rat-feeding fleas. Fleas also spread animal diseases such as cat and dog tapeworm and their biting can cause serious skin conditions such as dermatitis in animals.

It is hard to find a rose among the thorns when it comes to fleas, but there is one species that can be beneficial to humans. The European rabbit flea was introduced to Australia in 1966 to help spread myxomatosis, a disease used in the control of rabbits.

Fascinating facts

- Some fleas are very picky when it comes to eating. The appropriately named echidna flea feeds exclusively on the blood of echidnas.
- At the other end of the scale is the southern kiore flea, which has been found on 35 different animals including marsupials, rodents and domestic livestock.
- Fleas love rats! A close inspection of bush rats in Australia has found some animals with as many as 22 different species of flea feasting on their blood.
- The unusual translation of the word Siphonaptera comes from 'siphon' meaning tube or pipe, and 'aptera' meaning without wings. To loosely translate, they are wingless insects with tube-like mouths.
- Resilin is a special protein found in the legs of jumping insects, such as fleas. This rubber-like material stores energy,

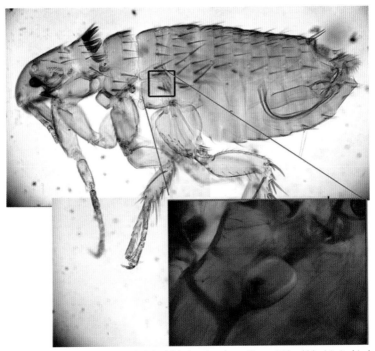

Close-up of a cat flea (*Ctenocephalides felis*) showing the resilin pad (dyed blue) in its hind leg. This rubber-like material stores energy, allowing fleas to rapidly leap great heights.
Source: Science Image.

allowing insects to suddenly leap great heights. Resilin is so efficient at storing and releasing energy that scientists have been studying its role in the bodies of jumping insects in the hope of creating a new generation of synthetic materials to be used in industry and medicine.

Learn more

In this book:

Cat fleas, p. 247

In other books:

Robinson WH (2005) *Urban Insects and Arachnids: A Handbook of Urban Entomology*. Cambridge University Press, Cambridge.
A detailed guide to cosmopolitan species of insects and arachnids, including fleas, which invade our homes and backyards.

Contains black and white drawings of commonly encountered species. ●●

An explanation of the above book ranking system can be found in Further Reading.

Online, type in:

Bubonic plague – also known as the black plague. Learn more about this flea-transmitted disease that killed an estimated 25 million people in Europe in the 14th century.

Flea jumping ability – high-speed video footage illustrates how fleas are able to jump up to 200 times their own body length.

Flea SEM – an images search will reveal scanning electron microscope (SEM) pictures of fleas, showing amazing details that are invisible to the naked eye.

Flies – Order Diptera

Summary – Flies

Order Diptera, meaning 'two wings'

152 244 species worldwide, 8000 species in Australia

Key features
- Only one pair of membranous wings present; hind wings reduced to tiny club-shaped knobs
- Legs slender, all equal in size
- Large compound eyes

Mosquitoes, midges, crane flies and their allies, Suborder Nematocera

Key features
- Antennae are longer than the thorax
- Bodies are usually long and slender

Giant mosquito (*Toxorynchites speciosus*).

House flies, blowflies, horse flies and their allies, Suborder Brachycera

Key features

- Antennae are shorter than the thorax and sometimes bristle-like
- Bodies are usually short and stout

Tachinid fly (*Rutilia* (*Donovanius*) sp.).

Don't confuse flies with:

Bees and wasps

Order Hymenoptera, see p. 162

Things in common

- Often similar in size and shape
- Membranous wings

Differences

- Bees and wasps have two pairs of wings, whereas flies have only one pair
- Bees and wasps have a narrow 'waist' where their thorax and abdomen meet, whereas flies usually have a thick 'waist'
- Bees and wasps often have antennae that are longer or thicker than those of flies

Bluebanded bee (*Amegilla* sp.).

Cicadas, leafhoppers, planthoppers and their allies

Order Hemiptera, Suborder Auchenorrhyncha, see p. 285

Things in common

- Often similar in size and shape
- Membranous wings
- Antennae bristle-like, like those seen in some flies

Differences

- Cicadas and hoppers have two pairs of wings held roof-like over their bodies, whereas flies have only one pair
- Flies often have much larger eyes than cicadas and hoppers, even taking up most of the head

Razor grinder cicada (*Henicopsaltria eydouxii*).

Description

Flies are among the most commonly encountered insects in our homes and gardens. We shoo them away from our food, swat at them as they attempt to suck our blood and inadvertently swallow them while jogging or cycling outdoors.

Flies do just that – they fly! – and it is a unique modification to their wings that makes them one of the most aeronautically gifted insects around. Unlike all other winged insects, which have four wings, flies have only two, with the hind wings reduced to a pair of tiny club-shaped knobs called halteres. These structures rapidly vibrate during flight and are what gives some flies their distinct whining sound as they zoom past. Halteres act like gyroscopes, helping flies to monitor their balance and adjust their steering as they fly. Their fore wings are usually clear and are responsible for propelling their bodies through the air.

Flies form one of the biggest Orders of insects. The group is very diverse: it includes mosquitoes, midges, moth flies, crane flies, fruit flies, blowflies, house flies, flesh flies, sand flies and hover flies. Robber flies have a wingspan of up to 75 mm; blowflies have short stout bodies in brilliant shades of metallic blues and greens; crane flies have extremely long slender bodies and legs; and midges are so small they are barely visible to the naked eye.

Flies have highly modified sucking, piercing or sponging mouthparts, depending on the types of food they eat. Their compound eyes are usually very large and, in some species, take up almost the entire head. Many flies also have three simple eyes (ocelli) on the crown of their head to aid their vision. Their legs are slender and covered in tiny, highly sensitive bristles. Many flies have pad-like structures located between the claws at the end of their feet, known as empodia (sing. empodium). As well as providing excellent grip, in some species they produce a sticky substance to help the fly adhere to slippery surfaces such as glass.

The antennae of flies vary greatly in structure and the Order Diptera is broken into two Suborders based on the appearance of these appendages. Suborder Nematocera includes the more primitive groups of flies, such as mosquitoes and crane flies. These slender flies have antennae that are long and thin in structure and may be adorned with plumes of fine hairs. Suborder Brachycera includes the more stoutly built flies, such as house flies and blowflies. Their antennae are much shorter, with some shrunk to a pair of tiny bristles.

All adult insects have a special sensory receptor known as a Johnston's organ located on their antennae, and this is used to the greatest extent in flies. This highly sensitive structure helps flies to monitor air vibrations, allowing them to judge flight speed, monitor the frequency of their wing beats and sense gravity. It is also used by males to detect the sound of passing females.

Diet and habitat

Flies can be found in any environment, from barren deserts to tropical rainforests. As most flies breed in moist habitats, we see the highest number of species in areas where there are lots of trees and water. Adult flies and their larvae live in quite different habitats and feed on different foods, varying even between parent and offspring within the same species.

As adult flies lack chewing jaws, they are restricted to a mostly liquid-based diet. Some are vegetarians, lapping up nectar from flowers and feasting on fungi, rotting fruit and decomposing vegetation. Others have more questionable palates, sponging up the moisture from dung, rubbish and decaying animal carcasses, or targeting the tears, sweat and mucus produced by living animals. Some, such as robber flies, are aerial predators, swooping on other flying insects and seizing them with their long legs, then stabbing them with a needle-like mouth and sucking them dry.

Many flies suck blood from other animals, including humans. Mosquitoes use a fine hypodermic syringe to siphon our blood. Horse flies, or March flies as they are also called, lack the surgical precision of their mosquito cousins. They use a stout rasping mouth to stab at their victim then sponge up the blood as it pools in the wound.

Maggots (the larvae of flies) are usually found in moist locations. Some burrow through mud, moist leaf litter, dung, compost or decomposing animal carcasses, using their hook-like mouthparts to extract plant and animal matter. Others, such as mosquito wrigglers, swim actively and are found in creeks, dams, ponds and marshes, using their bristle-like mouthparts to filter food from the water. Tachinid flies begin their life as parasites within the bodies of other insects such as grasshoppers and true bugs. They feast on the tissue of their host (usually killing it in the process) before leaving the carcass, pupating and changing into an adult fly.

Lifecycle
Courtship in flies is an elaborate process, often very similar to that of humans. Male fruit flies, for example, swarm at dusk in large groups known as leks, like humans gathering at a nightclub. They try desperately to attract the attention of passing females by emitting sounds (think of wolf-whistles) and by releasing wafts of attractive chemical signals known as pheromones (the fly equivalent of aftershave). Once a female shows interest, her affection is won through an elaborate dance-off, with carefully choreographed movements of the male's body, legs and wings.

Once fertilised, the female lays her eggs in a nice moist location. In mosquitoes, eggs are laid in rafts in the water; other flies inject their eggs into fruit, scatter them onto leaf litter or, in the case of parasitic flies, deposit them on or near the body of another insect.

Flies undergo complete metamorphosis, where the young do not resemble their winged parents in the slightest. Instead, they are pale legless wriggling larvae, usually referred to as maggots (see p. 68). They moult their skins to accommodate their increasing girth, much like a snake. When they have reached a certain size, the maggots often wriggle away from their larval habitat, looking for a drier location for the next stage of their lifecycle.

There they turn into a pupa, the stage in which the maggot transforms into an adult fly. Sometimes the pupa is enclosed within a silken bag (cocoon) for protection, in other species it remains naked. Some flies retain the last shed skin from their days as a maggot and use it as a protective capsule in which to pupate. This hardened brown barrel-shaped structure is known as a puparium. To emerge from this structure, some flies have an inflatable air-bag on their head; as it expands it pushes the top off the puparium. Once the fly has emerged it deflates the bag, tucking it neatly back into a special crevice on its forehead.

Defence
Due to their relatively small size, flies fall prey to many predators. Birds, lizards, frogs and spiders are the worst offenders, but flies also have to deal with the ravenous appetites of other insects. Dragonflies can often be seen swooping through the air, chasing mosquitoes as they swarm. Even other flies are foes – robber flies and longlegged flies are predators that happily feast on other types of flies.

Flies rely on their incredible speed and agility to outmanoeuvre attacks. Their enormous eyes and highly sensitive antennae act as an early warning system, alerting them to danger and allowing them to move out of reach in an instant.

Maggots also fall prey to predators. With no legs, it can be quite difficult to outrun an attacker, so fly larvae have developed other strategies. The larvae of hover flies live on plants, where they feast on aphids. Ants often stand guard over the aphids, feeding on the sugary secretions they produce. If an ant threatens a hover fly maggot, the larva bends over the ant and smothers it in a sticky droplet of saliva, rendering the attacker immobile. Some maggots, such as those of fruit flies, bend their bodies into a U-shape then suddenly contract their muscles, sling-shotting themselves out of harm's way.

Goodie or baddie?

With their bloodsucking tendencies, some flies have a negative impact on both humans and animals. Around the world, mosquitoes are responsible for the transmission of serious illnesses such as malaria, sleeping sickness, filariasis and yellow fever. Here in Australia, they cause outbreaks of Ross River fever, Murray Valley encephalitis and, in certain areas, dengue fever.

The Queensland and Mediterranean fruit fly lay their eggs in fruit crops and their hungry maggots burrow through our mangoes and cherries, turning them rotten and squishy. Flies also have a very high annoyance factor, whether it be house flies crawling on our food, bush flies landing on our mouths and eyes when outdoors, or the all-too-familiar whine of a mosquito at night.

However, some species out there are more friend than foe. Predators such as robber and hover flies and parasites such as tachinid flies help to keep insect populations under control, including those of many pest species. The nectar-feeding behaviour of other flies makes them good pollinators to have in the garden. And aquatic fly larvae, such as mosquito wrigglers, provide an important food source for many juvenile fish and crustaceans – something to think of next time you tuck into a seafood platter.

Fascinating facts

- Only female mosquitoes drink our blood as they need it to help develop their eggs. And they don't target only humans. The blood of dogs, cats, cattle, birds, lizards and even frogs is on the menu. Male mosquitoes obviously don't produce eggs and so they follow a vegetarian diet of nectar and plant secretions.

- Size does count, at least in the world of flies. Antlered flies are a species of fruit fly that occur in rainforests in the far northern tip of Australia. Female antlered flies lay their eggs in the decaying wood of a certain species of rainforest tree. Prime egg-laying sites are limited, so male flies seek out suitable locations and fiercely guard them. Males use the strange red 'antlers' on their heads (to which they owe their name) to protect their territory against rival males. When conflicts arise, they are not settled by the butting of heads, like their mammalian counterparts. Instead, they are solved in a much less brutish manner. The larger the male the bigger the antlers, which increases the odds of winning a physical battle. So rival males compare heights, stretching up onto their hind legs and raising their fore legs above their heads to see who is bigger. If a male does not measure up to his competitor, he leaves the fighting arena. However, if two males are evenly matched, a pushing contest ensues.

- Not all flies lay eggs. Bird flies (also known as louse flies or bat flies) are parasitic and feed on the blood of mammals and birds. Instead of the mother laying eggs, they hatch within her body and the larvae develop by feeding on special 'milk glands' within her abdomen, much like an infant child.

Once the larvae are fully grown, the mother deposits the maggots from her body onto a nearby substrate where they turn into pupae.

- Fly-strike is a serious condition affecting livestock such as sheep. Female blowflies do not lay eggs, but instead drop small maggots directly into wounds and crevices on the body of sheep. The maggots attack the living flesh and produce wounds that not only result in serious bacterial infections, but provide further sites for other blowflies to lay their maggots.
- The hairy maggot blowfly has an unusual family life, with individual females producing unisexual families. That means that throughout her life, a given female produces either all-male or all-female offspring. They are also unusual in that their maggots hunt the larvae of other flies. They get a good grip on their prey by wrapping their flexible bodies around it and use their hook-like mouthparts to consume their meal.
- Heard of glow worms, those beautiful shimmering creatures that adorn the walls and ceilings of certain caves? As they are the larvae of a type of fly known as a fungus gnat, they should really be called 'glow maggots'!
- Robber flies are large aerial predators which attack and eat other insects. Struggling prey can be quite bothersome to handle and a well-aimed kick may damage the delicate eyes and face of their captor. For this reason, robber flies have an impressive moustache of long hairs on their face to help cushion the blows.
- Warning! This is not for the squeamish, so you may want to put down that biscuit before you read on. Maggot therapy is a technique used by doctors to treat persistent wounds and infections that aren't responding to conventional treatment. Laboratory-reared maggots of certain blowfly species, bred under absolutely sterile conditions, are introduced to the wound, where they feast on dead infected tissue. They are surprisingly fussy eaters and leave healthy tissue virtually untouched, making them more efficient at removing dead tissue than even the most gifted surgeon. As they feed, they flood the wound with waste products which have amazing antibiotic properties, helping to disinfect and promote healing. Maggot therapy dates back to prehistoric times and gained much attention during the First World War, where wounded soldiers in the field were often brought back to hospitals with maggot-infested wounds. Maggot therapy is again gaining popularity and is used to fight infections in the wake of the ever-increasing resistance of bacteria to antibiotics.

Learn more

In this book:

Biting midges, p. 136
Bloodworms, p. 101
Crane flies, p. 113
Cucumber flies, p. 169
Flesh flies, p. 81
Horse flies, p. 137
Hover flies, pp. 133, 164, 184
Longlegged flies, p. 112
Mosquito adults, p. 135, wrigglers, p. 100
Moth flies, p. 83
Non-biting midges, pp. 101, 280
Snail parasitic blow flies, p. 287
Soldier flies, p. 102
Vinegar flies, p. 77

Flies play an important role in recycling nutrients. These blowflies (*Chrysomya* sp.) are laying their eggs onto the carcass of a rat, which will be fed upon by their maggot larvae. Source: K. Ebert.

This robber fly (*Ommatius* sp.) uses its stabbing mouthparts to feast on a small wasp.

These cucumber fly larvae (*Bactrocera cucumis*) are a major pest in crops such as papaya, zucchini and (pictured here) butternut squash. Source: D. Papacek.

A banana stalk fly (*Derocephalus angusticollis*) is attracted to the smell of decomposing fruit and vegetable scraps in a worm farm.

In other books:

Marshall SA (2012) *Flies: The Natural History and Diversity of Diptera.* Firefly Books, New York.
This book is beautifully photographed and contains detailed information about the different Families of flies that occur around the world. While the chapters contain lots of scientific names and terms, the descriptions as a whole are easy to read, interesting and very informative, making it a great book for anyone who wants to learn more about fly biology and classification. ●●

Waldbauer G (2003) *What Good are Bugs? Insects in the Web of Life.* Harvard University Press, London.
Interesting, easy to read information on the role of flies as disease vectors, pollinators of plants and waste recyclers, and more details on maggot therapy. ●●

An explanation of the above book ranking system can be found in Further Reading.

Online, type in:

Beyond Methuselah fly gene ageing – follow the links to a very interesting article involving fruit flies, cutting-edge genetic research and how it could help humans live longer.

Fly compound eyes – type this into an images search and you will have dozens of amazing close-up photographs of fly eyes staring back at you.

Fly feeding – learn about the unsavoury eating habits of house flies and watch videos of their amazing sponging mouthparts in action.

Maggots and murders – gain insight into the fascinating world of forensic entomology and learn how maggots can help determine time of death, if a body has been moved, or if a victim has been drugged.

Maggots time lapse – this will take you to a range of time-lapse videos of maggots devouring animal carcasses. Amazing stuff, but not for the squeamish!

Mosquito-borne diseases – learn more about diseases transmitted by mosquitoes, such as dengue fever, Ross River fever and Barmah Forest virus. Check to see which occur in your local area and read tips on how to avoid mosquito bites.

Mosquito emerging from pupa – watch mosquitoes emerge from their underwater pupae, without getting so much as a toe wet.

Swarming lake flies – watch amazing footage of huge clouds of lake flies, a type of non-biting midge, taking to the air in order to reproduce.

Grasshoppers and crickets – Order Orthoptera

Summary – Grasshoppers and crickets

Order Orthoptera, meaning 'straight wing'

23 616 species worldwide, 3000 species in Australia

Key features

- Elongate bodies
- Often well camouflaged
- Enlarged hind legs for jumping
- Leathery fore wings loosely cover membranous hind wings; wings sometimes absent
- Usually with large compound eyes
- Antennae long and thin (grasshoppers) or thread-like (crickets)

Grasshoppers and locusts, Suborder Caelifera

Key features

- Antennae long and thin
- Antennae usually shorter than body length

Longheaded grasshopper (*Acrida conica*). Source: A. Hiller.

Crickets and katydids, Suborder Ensifera

Synonym: Grylloptera

Key features

- Antennae thread-like
- Antennae usually longer than body length

Slender bush katydid (*Phaneroptera gracilis*).

Don't confuse grasshoppers and crickets with:

Praying mantids

Order Mantodea, see p. 250

Things in common

- Elongate bodies, often well camouflaged
- Leathery fore wings loosely cover membranous hind wings; wings sometimes absent
- Antennae long and thin
- Large compound eyes

Garden mantis (*Orthodera ministralis*).

Differences

- Praying mantids have enlarged **fore** legs lined with sharp spines for capturing prey, while grasshoppers and crickets have enlarged **hind** legs

Stick and leaf insects

Order Phasmatodea, see p. 259

Things in common

- Elongate bodies, often well camouflaged
- Leathery fore wings loosely cover membranous hind wings; wings sometimes absent
- Antennae long and thin

Differences

Red-winged stick insect (*Podacanthus viridiroseus*). The red colour of the wings is not visible when the wings are closed (as seen in this photo).

- Stick insects do not have enlarged hind legs for jumping. Instead all legs are long and slender, and used for walking

Description

Even the least bug-savvy person can usually identify a grasshopper or cricket – their large, powerful jumping legs make them easy to distinguish from other insects. The hind legs are greatly enlarged with huge bulging thighs, much like a chicken drumstick. It is the strong muscles housed within the hind legs that enable these insects to leap into the air. Proportionally, if humans were able to jump as well as a grasshopper or cricket, we could leap 90 m into the air (the equivalent of a 30-storey building) and around 150 m in distance (the length of an Aussie Rules football field), in a single jump!

Grasshoppers and crickets range in size from pygmy mole crickets that are only a few millimetres long, up to robust hedge grasshoppers and king crickets that can be as long as the palm of your hand. Most have large compound eyes, but these are absent in some species living in dark places such as caves or underground. A pair of strong biting mandibles allows them to chew through tough food. The adults usually have wings, consisting of a leathery fore wing (tegmina) and a fan-like hind wing, both of which fold neatly over the body while at rest.

Grasshoppers and crickets are further divided into two separate Suborders. Crickets (sometimes referred to as long-horned grasshoppers) belong to the Ensifera, a group that includes bush crickets, king crickets, raspy crickets, mole crickets and katydids. Grasshoppers (referred to as short-horned grasshoppers) are in the Suborder Caelifera, which also includes locusts.

Diet and habitat
Grasshoppers and crickets are found in a wide range of habitats, including rainforests, grassy plains, deserts, cold mountain ranges and rocky crevices and caves. They can turn up in odd places too – ant crickets live in the underground chambers of ant nests, feeding on secretions produced by the adult ants.

Grasshoppers and locusts are usually found on vegetation, either perched on the foliage of plants, resting on sticks and stems, or balanced on blades of grass. Others rest on the ground, basking in the sun. Some crickets are nocturnal, spending their days in tunnels and burrows underground and only venturing out once the sun has gone down. Raspy crickets make tiny shelters out of vegetation or soil woven together by silk, while mole crickets live in (and sing from) chambers deep within the soil.

Grasshoppers and locusts are strictly vegetarians, using their strong jaws to munch their way through foliage. Many are quite picky and feed from only a handful of plant species. Crickets and katydids have a reputation for being scavengers and eat leaves, fruit, flowers and seeds. Some species of katydids have a strict diet of nectar and pollen from flowers and others are predators, with spiky legs and chests to help capture their prey, which consists of smaller insects.

Lifecycle
Grasshoppers and crickets have a similar way of growing, but differ greatly in the way they attract mates and lay eggs.

Katydids and crickets are known for their beautiful courtship songs. Males produce a sound by rubbing a little file on their fore wing over a raised vein (known as a scraper) on their opposite wing. This is a bit like rubbing a bow across the strings of a violin. Katydids usually sing at night, crickets may sing day or night and often their calls are pitched at a frequency beyond our hearing capabilities. Once mated, females lay their eggs one at a time, although many hundreds of eggs may be laid in total. Some use a special sword-like blade at the end of their body (an ovipositor) to insert their eggs into stems and leaves; other species glue them to twigs or deposit them in the soil.

Grasshoppers in Australia are not known for their musical abilities and do not produce elaborate mating calls. Instead, males flick their legs, wave their antennae and perform little dances to signal their romantic intentions to females. Their eggs are laid in batches within little pods called oothecae, which are often bound together with a layer of hardened foam. Female grasshoppers and locusts have much shorter ovipositors than crickets and katydids, and usually deposit their eggs in shallow burrows in sandy soil.

When grasshoppers and crickets emerge from their eggs, they are known as nymphs and look like miniature wingless versions of their parents. They undergo incomplete metamorphosis, and their wings develop in sheaths on their backs called wing buds. With each moult of their skin, the wing buds get larger and become more visible until the nymph finally turns into a fully winged adult. You can sometimes find the shed skins of grasshoppers draped on your plants at home. Katydids are much tidier and dispose of their old skins by eating them.

Defence

Grasshoppers and crickets are large and abundant, making them a favourite food source of many birds, mammals, reptiles and invertebrates (e.g. spiders). Their camouflage is legendary in the animal kingdom, with many mimicking leaves, bark, sticks, pebbles, twigs and lichen to blend seamlessly into their surroundings. Many leaf-mimics concede that not all foliage is perfect and often have small brown spots and discolourations to make their disappearing act even more convincing.

If their enemies do happen to spot them, grasshoppers and crickets have more proactive measures to ensure they are not eaten. Large species bite with their sharp mandibles, others lash out with their spiky hind legs. Raspy crickets make scary hissing noises by rubbing their thighs against their body, and some grasshoppers gnash their mandibles together to make a loud squeak. King crickets and mole crickets can produce foul-smelling secretions from their anal glands – a surefire way to discourage a would-be attacker! Many brightly coloured species are advertising distasteful or poisonous chemicals within their bodies.

A quick exit is a handy trick to avoid being eaten. Grasshoppers and crickets have a special protein called resilin in their hind legs which helps with muscle elasticity, enabling them to leap large distances away from danger. If a bird is fast enough to seize one, limbs are often shed, like a lizard losing its tail, allowing the insect to escape the bird's grasp and flee.

Goodie or baddie?

Grasshoppers and crickets can be an annoying presence in our gardens. Grasshoppers chew large holes in ornamental plants and vegetables, and crickets and katydids eat the fruits and flowers on our shrubs. But they provide a vital food source for many animals and it is delightful to see a crested hawk swoop down from the trees, only to reappear moments later with a huge grasshopper clenched in its beak.

Here in Australia, swarms of locusts cause millions of dollars worth of damage to commercial crops, disrupt sporting games and other outdoor events and make driving and venturing outside almost impossible at times. Usually locusts live a solitary existence. But after large amounts of rain, the favourable conditions cause populations to explode and individuals start aggregating in large swarms and spread through the countryside in search of food. A typical locust swarm can contain over a billion individuals and cover tens of square kilometres.

Future Fido?

With their big expressive eyes and quirky camouflage, crickets and grasshoppers make interesting pets. Katydids and crickets are probably your best bet – their legs aren't as spiny as those of a grasshopper, making them easier to handle, and they eat a wider range of foods.

Fascinating facts

- The Nullarbor cave cricket is an opportunistic scavenger. Kestrels (small raptorial birds) sometimes build nests near the entrance of caves where these crickets reside. When the adult birds leave to forage, the crickets have been observed raiding the unprotected nest to feed on the tiny chicks.
- Chameleon grasshoppers are found only in high alpine regions in southern Australia and rarely live at altitudes lower than 1680 m. These amazing creatures change their colour (like their namesake) depending on the temperature. On a sunny day, males are blue. But on cloudy days when the temperature plummets, they turn black, allowing their chilly bodies to absorb as much heat as possible from solar radiation.
- Against the blackened landscape of a recent fire, a green grasshopper stands out like a sore thumb. But over the space of a generation some populations can darken their colour, helping them to better blend into their surroundings.
- The fireman grasshopper is found in northern Australia. It gets the name from its habit of sliding down grass stems, like a fireman going down a pole.

Learn more

In this book:

Ant-mimicking katydid nymphs, p. 180
Australian plague locusts, p. 143
Field crickets, p. 90
Hedge grasshoppers, p. 111
Katydids, p. 91
Kuranda spotted katydids, p. 252
Mole crickets, p. 89
Mottled katydids, p. 2
Nicsara katydids, p. 261
Vegetable grasshoppers, p. 124

In other books:

Rentz D (2012) *A Guide to the Katydids of Australia*. CSIRO Publishing, Melbourne. A beautifully photographed field guide to Australian katydids. ●●
Rentz D (1996) *Grasshopper Country: The Abundant Orthopteroid Insects of Australia*. UNSW Press, Sydney.

What is the difference between a grasshopper and a locust?

To look at, grasshoppers and locusts are the same – the difference between them comes down to their behaviour. Locusts may or may not form large migrating swarms, whereas grasshoppers generally do not. This distinction between the two is blurry – some grasshoppers come together in loose groups (although not big enough to define as a 'swarm') and the names 'grasshopper' and 'locust' are used interchangeably.

This bark cricket (*Zaclotathra oligoneura*) is almost invisible against the trunk of a tree.

This juvenile spotted bandwing grasshopper (*Qualetta maculata*) blends in well with the red sands of western Queensland.

The white spots on this juvenile spotted predatory katydid (*Chlorobalius leucoviridis*) reflect sunlight, presumably to help it blend into the dense foliage where it shelters during the day.

Detailed information on the different families of grasshoppers and crickets in Australia, including identification keys which can be used (with the aid of a microscope/magnifier) to identify various Families, Subfamilies and Tribes. Colour plates illustrate some commonly encountered species. ●●

Rentz DCF, Lewis RC, Su YN and Upton MSD (2003) *A Guide to Australian Grasshoppers and Locusts*. Natural History Publications, Borneau, Kota Kinabalu.

A beautifully photographed field guide to Australian grasshoppers and locusts, containing close-up photographs and distribution maps for each species and a handy chapter on identifying nymphs of different species. ●●

An explanation of the above book ranking system can be found in Further Reading.

Online, type in:

Australian Plague Locust Commission – for information on locust plagues in Australia.

Cooloola monster – this bizarre cricket, found in 1976 in Cooloola National Park, Queensland, was so unique that it needed to be classified into a Family of its own.

Leichhardt's grasshopper – as elusive as it is beautiful, this striking orange and blue grasshopper was discovered by German explorer Ludwig Leichhardt in 1845.

Locust swarm – do an images search to see photographs of locust swarms wreaking havoc on people, animals and crops.

Weta insect – native to New Zealand and closely related to Australian king crickets, these massive crickets are one of the heaviest insects in the world.

Lacewings – Order Neuroptera

Summary – Lacewings

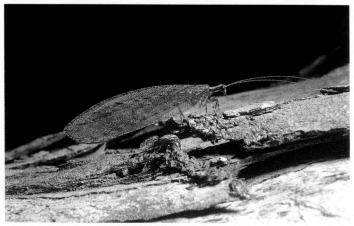

Lacewing (*Stenosmylus* sp.). Source: Gooderham and Tsyrlin (2002).

Order Neuroptera, meaning 'nerve wing'

Synonym: Planipennia

5704 species worldwide, 600 species in Australia

Key features

- Elongate slender bodies
- Two pairs of similarly sized, blade-like membranous wings, with networks of tiny veins
- Wings held roof-like over the body or out away from the body at rest
- Large compound eyes
- Antennae long and thin, sometimes clubbed

Don't confuse lacewings with:
Dragonflies and damselflies
Order Odonata, see p. 212

Things in common

- Elongate slender bodies
- Two pairs of similarly sized, blade-like membranous wings, with networks of tiny veins
- Large compound eyes

Differences

- Dragonflies and damselflies have tiny bristle-like antennae almost invisible to the naked eye, whereas lacewings have antennae that are long and thin, or clubbed
- Dragonflies and damselflies do not hold their wings roof-like over their bodies while at rest

Common bluetail damselfly (*Ischnura heterosticta*).

Praying mantids

Order Mantodea, see p. 250

Things in common

- Netwinged mantids are an unusual type of praying mantis that have lacy-looking wings held roof-like over their body at rest, much like a lacewing
- Mantis flies (see photo on p. 253) are a special group of lacewings that have large fore legs lined with sharp spines, similar to those of praying mantids

Netwinged mantis (*Neomantis australis*). Source: A. Hiller.

Differences

- Although they may look lacy, the fore wings of praying mantids are actually quite leathery and opaque and they do not have as many veins as the wings of lacewings
- The eyes of mantis flies are very round and globular compared to the eyes of praying mantids

Cicadas, leafhoppers, planthoppers and their allies

Order Hemiptera, Suborder Auchenorrhyncha, see p. 285

Things in common

- Membranous wings held roof-like over the body while at rest

Differences

- Cicadas and hoppers have broader bodies than lacewings and their antennae are bristle-like and barely visible
- The wings of cicadas and hoppers have fewer veins than the wings of lacewings

Brown bunyip cicada (*Tamasa tristigma*). Source: K. Ebert.

Moths

Order Lepidoptera, see p. 197

Things in common
- Moth lacewings (see p. 199) are a type of lacewing that have wings clothed in tiny hairs, giving them a moth-like appearance

Differences
- Moths have heads and bodies covered in furry scales, whereas moth lacewings do not (they may have short hairs on their bodies and legs, but these do not give them a furry appearance)
- Moths have wings covered in flattened scales whereas moth lacewings have wings that are hairy
- Moths' wings have fewer veins than the wings of lacewings

Pasture webworm (*Hednota grammellus*). Source: T. Daley.

Description

Lacewings are easily recognised by their two pairs of large gauzy wings that are roughly equal in size. Their blade-like wings are criss-crossed with a dense network of tiny veins, giving them a delicate lacy appearance. While they closely resemble dragonflies and damselflies, lacewings are no match when it comes to aerial prowess, with most species achieving little more than a slow flutter as they move through the air. While at rest, their wings are held like a little roof or an upside-down V over their bodies, a trait that sets them apart from many other insects.

There are many different types of lacewings, such as green and brown lacewings, mantis flies, moth lacewings, antlions and owlflies. They range in size from tiny insects with a wingspan of less than 5 mm, through to large specimens with wings reaching up to 150 mm. Their wings may be shiny and transparent, covered in colourful patterns or clothed in tiny hairs.

Lacewings have large spherical compound eyes and long thin antennae that may be thickened into a little club at the end. They have strong biting mouthparts and slender legs for walking.

Diet and habitat

Lacewings favour tropical environments but can also be found in dry desert habitats, open forests and along the edges of cold mountain streams. While some brightly coloured species are active by day, most lacewings fly at dusk or are nocturnal. Many species are attracted to bright lights and so we often encounter lacewings around our windows, doors and ceilings at night.

Adult lacewings prey on small, soft-bodied insects and some species also consume nectar and pollen from flowers. But their larvae feed exclusively on other animals and have a reputation for being highly skilled, voracious predators. Armed with a pair of hollow, razor-sharp jaws, lacewing larvae impale their prey and suck it dry. Green and brown lacewing larvae prowl the stems and leaves of plants, pouncing on sap-sucking insects such as aphids and mealybugs. Antlions prefer to lay a trap, digging conical pits in sandy soils where they wait hidden at the bottom until an unsuspecting ant blunders into their open jaws. There are even aquatic lacewing larvae known as spongeflies which, as the name suggests, devour freshwater sponges.

Lifecycle
To attract the interest of a female, some male lacewings use the insect equivalent of cologne – a chemical odour (pheromone) secreted from special glands on the body. Others entice females musically, with some species tapping their abdomen to produce courtship songs.

A female may lay a few eggs or a few hundred, depending on the species. These oval-shaped eggs may be scattered on the ground or cemented to leaves and other surfaces. Some species suspend each egg on top of a tall strand of silk, placing them out of reach of predators such as ants and even each other – so voracious is the appetite of a newly hatched larva that it devours its unhatched siblings if given the chance.

Lacewings undergo complete metamorphosis, growing and changing through a process similar to that of a butterfly. The larva usually moults its skin three times before spinning a small cocoon of silk, from which it eventually emerges as an adult. Depending on the species, the lifecycle may take several weeks to a few years.

Defence
As with most insects, lacewings fall prey to larger animals such as birds, spiders and frogs. Bats prey on nocturnal species and some lacewings drop suddenly to the ground to dodge the ultrasonic calls bats use to locate their prey. Day-flying lacewings, such as some mantis flies, mimic the bright warning colouration of dangerous wasps to deter enemies.

When handled, some green lacewings release a foul-smelling scent from their body, earning them the unflattering moniker 'stink flies'. If you have ever been on the receiving end of this pungent gas you know that no amount of frantic handwashing will budge it – you end up less popular than usual with friends and family for the next few hours!

Lacewing larvae are masters of camouflage. They pile up debris on their backs, which may include the sucked-out remains of their last meal, and hold it in place with long curved hairs on their body. This makeshift shelter not only confuses would-be predators, it helps the larvae to sneak up on their prey when hunting.

Goodie or baddie?
Lacewings are undoubtedly good guys, helping us to combat insect pests in our gardens and commercial crops. Both adults and larvae are important natural enemies of plant-sucking nasties such as psyllids, aphids, scale insects, mites and mealybugs. The eggs and caterpillars of pest moths and butterflies are also targeted. There are several companies in Australia that breed populations of lacewings for mass release

into crops, allowing farmers to manage pest populations while minimising the use of harmful insecticides.

Fascinating facts

- Lacewing larvae construct a silken cocoon, much like that of a moth, inside which they transform into an adult. However, instead of producing silk from glands in their mouth (like caterpillars), lacewings unravel the silken strands through their anus!
- With their sleek bodies, speed, powerful strength and large protruding jaws, the larvae of brown lacewings are sometimes referred to as 'insect crocodiles'.
- The larvae of mantis flies are specialised parasites on spiders. The newly hatched lacewing larva hitches a ride on an unsuspecting female spider: when the time comes for the spider to lay her eggs, it sneaks inside the egg sac to feast on the contents.
- Green and brown lacewing larvae are very effective at controlling aphid populations, with a single larva capable of eating 60 aphids in an hour.

Learn more

In this book:

Antlions, p. 153
Blue eyes lacewings, p. 214
Green lacewings, p. 92
Lacewing eggs and larvae, p. 159
Mantis flies, pp. 252–253
Moth lacewings, p. 199

In other books:

Llewellyn R (2002) *The Good Bug Book*. Integrated Pest Management Pty Ltd, Australia.

A green lacewing larva (*Mallada signatus*) uses its piercing mouthparts to feast on black citrus aphids (*Toxoptera citricida*). The larva piles up the carcasses of its victims on its back to disguise itself while hunting. Source: D. Papacek.

This owlfly (*Suhpalacsa* cf. *lyriformis*) has distinctly clubbed antennae. Source: John Tann, Attribution 2.0 Generic.

Great information on insects, such as lacewings, used in the biological control of crop pests. ⬥

An explanation of the above book ranking system can be found in Further Reading.

Online, type in:

Antlion's death trap – watch great footage of an antlion constructing its pit and capturing prey.

Green lacewing biological control Australia – learn about the use of these beneficial insects in the control of pests such as aphids.

Lacewing larvae – do an images search to see examples of these formidable predators.

Mantidfly – do an images search to see this remarkable lacewing that features raptorial fore legs for capturing prey. There are even some individuals that mimic the bright colours of wasps.

Lice – Order Phthiraptera

Summary – Lice

Order Phthiraptera, meaning 'lice without wings'

Synonym: Psocodea (currently valid)

5024 species worldwide, 465 species in Australia

Key features

- Tiny, often brown
- Body dorso-ventrally flattened (short and flat)
- Never have wings
- Legs short and stout with large claws on the ends

Head louse (*Pediculus humanus capitus*). Source: Gilles San Martin, Attribution–ShareAlike 2.0 Generic.

Don't confuse lice with:

Fleas

Order Siphonaptera, see p. 221

Things in common

- Tiny, usually brown
- Never have wings
- Both fleas and lice are bloodsucking parasites of other animals

Differences
- Fleas are laterally compressed (tall and skinny) whereas lice are short and flat
- Fleas have longer skinnier legs than lice

Cat flea (*Ctenocephalides felis*). Source: J. Dorey.

Booklice

Order Psocoptera*, see p. 194

Things in common

- Tiny, often brown
- Many booklice lack wings

Differences

- Booklice have long thread-like antennae whereas lice have antennae that are short and stubby
- Booklice do not have large claws at the end of their legs

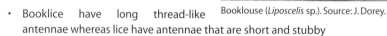

Booklouse (*Liposcelis* sp.). Source: J. Dorey.

*Lice and booklice have recently been grouped together under the Order Psocodea, however, this book uses their old Order names (the reason for which is explained in Chapter 2, p. 19).

Description

Lice are highly specialised insects that spend their entire lives as parasites on the bodies of other animals. While individual species can vary greatly in size and shape, they are all small (less than 10 mm long), wingless and usually drab coloured. Their bodies are extremely flattened and their legs end in strong claws, both modifications that enable them to cling to the body of their host.

It is usually dark living among feathers or fur, so lice have reduced eyes or none at all. They instead rely on their short antennae, which may be tucked into little grooves on the sides of their head, as their main sensory organs. The surface of their antennae is covered in tiny sensory hairs called setae that help them navigate around the animal's body and sense the smell and temperature of their host.

Lice can be divided into two categories, biting or sucking, depending on the appearance of their mouthparts. Biting lice have strong jaws used for nipping and chewing on the feathers or skin of their host. They can be found on a variety of birds and other animals. Sucking lice have sharp piercing mouthparts and feed exclusively on the blood of mammals.

Diet and habitat

All stages of the lice lifecycle are found on the bodies of warm-blooded animals. As a collective group, lice feed on many types of animals, including rats and mice, native animals such as kangaroos, wallabies and dingoes, domestic animals such as dogs, cats, guinea-pigs, rabbits, horses, goats and cows, and all species of birds. However, individual species of lice are very loyal to a certain type of animal, and even a particular area on the animal's body.

The human louse, for example, is known to comprise two sibling species. Each lives on a different part of the human body. Head lice (*Pediculus humanus capitus*) are found on the scalp and hair, and body lice (*Pediculus humanus humanus*) in our clothing and nether regions. The genetics of these two types suggests that the body louse evolved around 107 000 years ago – around the same time primitive humans started wearing clothes, under which the lice shelter.

Lice move from animal to animal through physical contact. This may occur during mating or the nursing of young, and in communal sleeping places such as roosts and burrows. In humans, head lice spread like wildfire in schools and childcare centres, where head-to-head contact often occurs as children play. Unlike their bloodsucking comrades, the fleas, lice cannot survive long off the body of their host.

Lifecycle

In nature, female lice often outnumber males and therefore finding a potential mate can be difficult. For this reason, some species of lice undergo parthenogenesis, a process by which fertile eggs are produced without pairing with a male. Female lice fasten their eggs (commonly referred to as 'nits') to the feathers or hair of their host with a cement-like glue. Females usually lay a few eggs each day, with some species laying several hundred during their lifetime. To emerge from the egg, the hatchling louse gulps mouthfuls of air into its gut, blowing up like a balloon until the pressure causes the top of the egg to rupture. The human louse goes one better, with sharp blades on the front of its head to help pierce through its eggshell.

Lice go through incomplete metamorphosis and so the hatchling nymph closely resembles its parents, but is smaller and paler. It gradually grows larger and darker through a series of moults. The whole lifecycle takes between two and six weeks, depending on the species.

Defence

Lice are generally not preyed upon by other animals – they are tiny and secretive. Their biggest enemy is their own host animal, which frantically scratches, grooms or preens its body to rid itself of its annoying tenants. Many animals have dust baths, rolling in the dirt to try to suffocate the lice that infest their bodies. Some species of birds keep lice populations under control through an amazing process known as *anting*. Birds voluntarily sit on ant nests, allowing the insects to swarm over their bodies and roam among their feathers, looking for lice to consume. Other birds are not nearly as patient – they pick up ants in their beaks (usually crushing them in the process) and dab them over their plumage, much like a woman anointing herself with perfume. When applied to the feathers, the noxious formic acid produced in the ant's body discourages infestations of lice.

Goodie or baddie?

Human lice can cause nasty skin reactions, which can turn into secondary infections through vigorous scratching. In areas where human contact is high (e.g. schools, childcare centres and family homes) infestations of head lice can be ongoing and lice rapidly become resistant to the chemical shampoos and concoctions we use to control them.

As with most bloodsucking insects, lice can transmit diseases to humans and animals. Human lice can transmit epidemic typhus, which thankfully has been absent from Australia for quite some time. The dogbiting louse is an intermediate host for dog tapeworm. Lice can also cause substantial economic loss in dairy, poultry, cattle and other agricultural industries. They cause skin conditions such as dermatitis, and heavy infestations can reduce the overall health of animals.

One louse that does more good than harm is the rabbit louse, which can facilitate the spread of myxomatosis, a disease used to control rabbit populations in Australia.

Fascinating facts

- Not even aquatic animals are spared the annoyance of lice. Fur seals, sea lions and leopard seals can all be plagued by infestations of sucking lice.
- 'Lice' and 'louse' is like 'mice' and 'mouse'. The word 'lice' is used for many; 'louse' is for a single individual.
- 3800 lice were once removed from the body of one unfortunate (and presumably very itchy) human.
- As they have no wings, lice are unable to fly in search of hosts. Instead they hitch a ride with someone who can. Bird flies (also known as louse flies) are a strange type of fly that lives alongside lice as parasites on the bodies of birds. When a louse fancies a change of scenery, it crawls onto the body of the fly and hangs on tightly. When the fly lands on a new bird, the louse disembarks. Each fly usually carries one or two passengers at a time, but one specimen (presumably exhausted!) was observed with 31 lice clinging to its body.
- Lice are tough. One species can be found on Weddell seals in Antarctica, clinging to the hairs on the tail and hind

flippers. These lice are regularly exposed to waters as cold as −2°C as the seals roam the frigid waters in search of food.

Learn more

In this book:

Bird lice, pp. 195, 222

In other books:

Waldbauer G (2003) *What Good are Bugs? Insects in the Web of Life*. Harvard University Press, London.
Interesting, easy-to-read information on the impact lice have on humans and animals, their role as disease vectors, and more details on anting and the differences between biting and sucking lice. ●●

An explanation of the above book ranking system can be found in Further Reading.

Online, type in:

Head lice myths – figure out fact from fiction by reading these myth-busting articles.

Lice human evolution – interesting articles on how the study of bloodsucking parasites such as lice can help us unlock our evolutionary history, such as when humans started wearing clothes and how much we interacted with other primates.

Lice SEM – an images search will reveal scanning electron microscope (SEM) pictures of lice, showing amazing details that are invisible to the naked eye.

Praying mantids – Order Mantodea

Summary – Praying mantids

Burying mantis (*Sphodropoda tristis*).

INSECT ORDERS 251

Order Mantodea, meaning 'prophet'

Synonym: Dictyoptera

2384 species worldwide, 160 species in Australia

Key features

- Elongate bodies with long slender legs
- Often well camouflaged
- Fore legs enlarged and lined with sharp spines
- Leathery fore wings loosely cover membranous hind wings; wings sometimes absent
- Triangular head with large compound eyes
- Antennae long and thin

Don't confuse praying mantids with:

Stick and leaf insects

Order Phasmatodea, see p. 259

Things in common

- Elongate bodies with long slender legs

Spiny leaf insect (*Extatosoma tiaratum*).

- Often well camouflaged
- Leathery fore wings loosely cover membranous hind wings; wings sometimes absent

Differences

- Stick insects have oval-shaped heads with tiny beady eyes, whereas praying mantids have triangular-shaped heads with huge eyes that take up much of the head's surface

Grasshoppers, crickets and katydids

Order Orthoptera, see p. 233

Things in common

- Elongate bodies
- Often well camouflaged
- Leathery fore wings loosely cover membranous hind wings; wings sometimes absent

Kuranda spotted katydid (*Ephippitytha kuranda*).

Differences

- Grasshoppers, crickets and katydids have enlarged hind legs used for jumping. Praying mantids have enlarged fore legs with sharp spines and slender walking legs along the rest of their body

Mantis flies

Order Neuroptera, see p. 241

Things in common

- Mantis flies are a type of lacewing that have enlarged fore legs with sharp spines for capturing prey
- Large eyes

Differences

- Mantis flies have transparent wings criss-crossed with a network of tiny veins, held roof-like over their bodies at rest. Praying mantids have fore wings that are leathery and opaque, with not as many veins, and usually fold their wings flat over their abdomens

Mantis fly (*Campion* sp.). Source: K. Ellingsen.

Description

With their long slender bodies and limbs, praying mantids are often confused with stick insects, grasshoppers and crickets. But a quick glance at their head easily distinguishes them from other leaf- or stick-mimicking insects. Praying mantids have a very alien-like head, triangular in shape and featuring enormous compound eyes. Their amazing eyesight, sensitive antennae and highly mobile head helps them to sense and stalk their prey. Praying mantids have strong enlarged fore legs lined with razor-sharp spines. These raptorial limbs snap closed with lightning speed, impaling prey in their vice-like grip.

Praying mantids have leathery fore wings (tegmina) which fold over clear, fan-like hind wings. Male praying mantids are generally good fliers, while many females have wings that are shortened or missing altogether. The females of some species, despite having long, well-developed wings, have large heavy abdomens which make flying impossible.

Diet and habitat

Praying mantids are strictly carnivorous, attacking and eating almost any animal small enough for them to overpower. While their diet mainly consists of other insects, larger mantids have been known to occasionally dine on frogs, lizards and even small birds!

Praying mantids are found in habitats where they can stalk potential prey. Their green or brown colouration helps them to stake out tall grasses, shrubs and tree canopies. There are even brightly coloured species that vanish among the petals of flowers, leaping out to attack thirsty bees or butterflies that fly in for nectar.

Lifecycle

With the possible exception of dung beetles and fly maggots, there is no worse job in the insect world than that of the male praying mantis. In the pursuit of love, the male suitor is sometimes eaten by his greedy girlfriend. As he mates with her, she can attack – lopping off his head so he can't bite back and eating his insides. But all is fair in love and war and this nutritious meal helps nourish her during the egg-laying process.

The eggs are laid in an ootheca, a spongy case that looks very much like a brown marshmallow. These cases are usually glued to sticks, leaves and grass although some species deposit them on the ground. An ootheca can contain between 10 and 400 eggs, depending on the species of mantis that produces it.

Praying mantids grow through incomplete metamorphosis, with the tiny nymphs closely resembling the adults. They feed on soft-bodied insects until they are big enough to catch larger prey, and grow by moulting their skin several times.

Defence

As well as helping them stalk their prey, praying mantids use camouflage to avoid becoming a larger animal's lunch. They can mimic bark, sticks, leaves, grass and flowers, often swaying their bodies gently in the breeze to make their disappearing act even more convincing.

If confronted by a would-be attacker, mantids can drop suddenly to the ground, run away or take to the air. Some species flash brightly coloured wings, spit foul-tasting fluid from their mouth or lash out with their spiky fore legs.

Goodie or baddie?

Having praying mantids in your garden ensures you will have one of the best veggie patches in town. These pint-sized assassins devour grasshoppers, plant-sucking bugs, moths and other nasties around your backyard, making them a great alternative to bug spray.

Future Fido?

Praying mantids make very interesting pets, especially if you watch them stalk and capture their food. An endearing feature are their pseudopupils – tiny black spots on their eyes that change position as the mantis moves its head, giving the impression they are watching you. To keep a praying mantis you need a steady source of suitably sized live insects, as they rarely touch dead food items. And most importantly, keep only one mantis per enclosure – being avid cannibals makes them terrible roommates!

Fascinating facts

- A giant green mantis was observed in northern Australia feasting on a green tree frog that weighed 25 g – around 3.5 times its own bodyweight!
- Praying mantids have a specialised hearing organ between their hind legs, allowing them to detect the ultrasonic calls used by bats to locate their prey. This provides mantids with an early warning system, giving them time to flee or drop to the ground before the bat can swoop.
- You will find the names of these insects spelt in different ways – **pray**ing (the way they hold their fore legs in a prayer-like posture), **prey**ing (referring to their hunting lifestyle) and either man**tis** or man**tid** (which can be used interchangeably). All spellings are considered to be correct.

Learn more

In this book:

    Garden mantids, p. 235
    Netwinged mantids, p. 242

This tiny praying mantis (*Calofulcinia* sp.) can be found hunting on tree trunks, where it mimics moss and lichen. Source: A. Hiller.

This unfortunate male giant green mantis (*Heirodula majuscula*) is decapitated and devoured after mating with his peckish partner. Source: A. Hiller.

Oothecae, pp. 157–159
Stick mantids, p. 260

In other books:

New TR (1991) *Insects as Predators.* UNSW Press, Sydney.
A very detailed look into how predators such as praying mantids locate, capture and consume their prey. ●●●
Rentz D (1996) *Grasshopper Country: The abundant Orthopteroid Insects of Australia.* UNSW Press, Sydney.
Beautiful colour photographs and detailed information on the different Families of praying mantids in Australia. ●●

An explanation of the above book ranking system can be found in Further Reading.

Online, type in:

5 praying mantis species that will blow your mind – this will take you to info and photographs of some of the world's most beautiful and bizarre praying mantids.
Female praying mantis eats male after mating – do a video or images search to see dozens of examples of a phenomenon known as sexual cannibalism.
Flower mantis – an image search will result in pictures, such as the devil's flower mantis, that look almost too spectacular to believe.

Praying mantis 3D glasses – scientists have created the world's smallest pair of 3D glasses and fitted them to the face of a praying mantis, to better understand their vision and depth perception.

Praying mantis hunting – heaps of photos, videos and information on how these insects hunt and consume their prey.

Silverfish – Order Thysanura

Summary – Silverfish

Order Thysanura, meaning 'fringe tail'

Synonym: Zygentoma (currently valid)

527 species worldwide, 36 species in Australia

Key features

- Small, carrot-shaped, flattened bodies
- Body usually covered in shiny silvery or grey scales
- Never have wings
- Eyes reduced or absent
- Antennae long and thin
- Three long, thread-like tails at the end of the abdomen

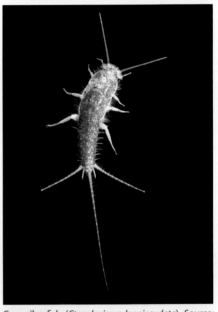

Gray silverfish (*Ctenolepisma longicaudata*). Source: Science Image.

Don't confuse silverfish with:

Cockroaches

Order Blattodea, see p. 207

Things in common

- Flattened bodies
- Some cockroaches are wingless

- Both cockroaches and silverfish are fast runners, frequently found in our homes

Differences

- Cockroaches are usually brown and have two short finger-like cerci at the end of their abdomen. Silverfish are usually silver or grey and have three much longer tails at the end of their bodies
- Cockroaches have large compound eyes and much longer antennae than silverfish

American cockroach nymph (*Periplaneta americana*).

Description

Silverfish get their name from the grey or silver overlapping scales that adorn their flattened bodies. They are quite small, usually no bigger than 25 mm long, and are shaped like a carrot with their bodies tapering towards the bottom end. A pair of long thin antennae adorn the top of their head and three long, thread-like, equally sized tails are found at the end of the abdomen.

Silverfish have chewing mouthparts and tiny reduced eyes that are sometimes missing entirely. Their slender legs are used for running and they are extremely agile. In fact, we know relatively little about silverfish – scientists have a hard time studying them because they run so fast when disturbed.

Silverfish are regarded as one of the most primitive insects on our planet. They never grow any wings, even as adults. Indeed, insects in this group have never had wings, a phenomenon which is unusual in the world of insects (other wingless insects, such as fleas, show evidence of having evolved from winged ancestors).

Diet and habitat

Silverfish love to shelter in small dark places such as under bark, among leaf litter and in caves. Some species lurk in the corridors of ant and termite colonies, while others take up residence in the bottom of bird nests. Unfortunately, silverfish also like to share our homes and can be found hiding in our cupboards and sinks.

Silverfish have a pretty broad diet and feed mostly on plant-based materials in the wild. They have the amazing ability to produce special enzymes to help them break down the tough cellulose in plant tissue, something that most insects are unable to do without the help of special organisms living in their gut. This robust digestive system enables them to munch through paper, glue, linen, wallpaper and book bindings in our homes.

Lifecycle

Male silverfish bundle up their sperm into little packages called spermatophores, which are deposited on the ground then picked up by the female. The females glue their small, oval-shaped eggs onto the ground or other surfaces such as paper.

Silverfish have an unusual lifecycle called simple metamorphosis (see pp. 21–22). Because these primitive insects don't have the burden of growing wings, their basic appearance remains more or less unchanged throughout their life and they simply grow larger as they shed their skins. They moult at least 10 times before they mature and can live for up to four years – an impressively long time for a small insect.

Defence

Have you ever tried to catch a silverfish? The moment they are uncovered from their hiding places they shoot off with lightning speed and agility. If you are fast enough to get hold of one, the slippery scales on its body allows it to wriggle out of your grip. These tiny escape artists are experts at thwarting would-be predators such as spiders.

Goodie or baddie?

Silverfish are more annoying than harmful. They do not spread any nasty germs or diseases and are harmless to humans. But they love munching little holes in our paper and books, making them an unwelcome visitor in our homes.

Fascinating facts

- Feeling thirsty? Silverfish can survive in dry environments by absorbing water from the atmosphere through the walls of their rectum. Basically, they can drink from their bottoms!
- Silverfish have the unusual ability to reproduce before they are fully grown. After reaching sexual maturity at two to three months of age, some silverfish mate and produce a batch of eggs between each of their moults. Since moulting will continue throughout their long lives, they have the potential for many offspring.

Learn more

In other books:

Robinson WH (2005) *Urban Insects and Arachnids: A Handbook of Urban Entomology.* Cambridge University Press, Cambridge.

A detailed guide to cosmopolitan species of insects and arachnids that invade our homes and backyards. Contains black and white drawings of commonly encountered species. ●●

An explanation of the above book ranking system can be found in Further Reading.

Online, type in:

Silverfish damage to paper – type this into an image search to see how destructive their feeding can be.

Silverfish lifecycle – learn more about the primitive and unique way in which they grow and reproduce.

Stick and leaf insects – Order Phasmatodea

Summary – Stick and leaf insects

Order Phasmatodea, meaning 'apparition'

Synonyms: Phasmatoptera, Cheleutoptera, Phasmida (currently valid)

2853 species worldwide, 150 species in Australia

Key features
- Extremely elongate bodies with long slender legs
- Often well camouflaged
- Leathery fore wings loosely cover membranous hind wings; wings sometimes absent
- Oval-shaped heads
- Compound eyes quite small
- Antennae long and thin

Confused winged stick insect (*Sipyloidea caeca*).

Don't confuse stick and leaf insects with:

Praying mantids

Order Mantodea, see p. 250

Things in common
- Elongate bodies with long slender legs
- Often well camouflaged
- Leathery fore wings loosely cover membranous hind wings; wings sometimes absent

Stick mantis nymph (*Archimantis latistyla*).

Differences

- Praying mantids have triangular-shaped heads and very large compound eyes, whereas stick and leaf insects have oval-shaped heads and much smaller eyes
- Praying mantids have enlarged fore legs lined with sharp spines, stick and leaf insects do not

Grasshoppers, crickets and katydids

Order Orthoptera, see p. 233

Things in common

- Elongate bodies with long slender legs
- Often well camouflaged
- Leathery fore wings loosely cover membranous hind wings; wings sometimes absent

Differences

- Grasshoppers, crickets and katydids have enlarged hind legs used for jumping, stick and leaf insects do not

INSECT ORDERS 261

Nicsara katydid (*Nicsara* sp.).

Description

Stick insects are among the largest insects in Australia. They can be up to 30 cm from head to tail – that's as long as a ruler! They have long slender bodies and limbs, often resembling sticks or leaves (hence their name). True leaf insects belong in a separate Family from stick insects (although both belong to the Order Phasmatodea), are quite rare and are usually found only around the far northern tip of Australia. For this reason, members of this Order are referred to here as 'stick insects' rather than 'stick and leaf insects' (although their biologies are more or less the same).

Stick insects have oval-shaped heads, with small beady eyes and strong chewing mandibles. Their long thin antennae are covered in tiny sensitive hairs called setae, which help them to detect their surroundings.

Stick insects have long legs used for walking, with a pair of claws and a suction cup at the end of each foot to help them climb. The fore wings form small flaps (tegmina) that fold over the large, fan-like hind wings. Many species have no wings at all.

Diet and habitat

Stick insects can be found throughout Australia, although they are most abundant in warmer climates. They live in rainforests, grasslands and open forests, usually clinging to the foliage of trees and shrubs. To find a stick insect in its natural habitat is quite a rare occurrence, as their amazing camouflage makes them almost impossible to locate. We most commonly see stick insects on windy days or after storms, when they are blown from trees and are easily spotted on our lawn, car roof or screen door.

Stick insects are devoted vegetarians, with both adults and juveniles feasting on the leaves and young stems of trees and shrubs. They usually target native plants such as eucalypts and wattles, but can also be found on a variety of other plants, including rose bushes, raspberry vines and guava trees.

Lifecycle

Male stick insects are usually smaller and much thinner than their rather plump girlfriends, whose abdomens become swollen with hundreds of eggs. When you look exactly like a leaf or stick, it can be very hard for a potential mate to spot you in a tree. For this reason, female stick insects produce enticing perfumes known as pheromones and release them from their bodies. The males use their long antennae to close in on these chemical cues, allowing them to locate the female among the foliage.

Upon mating, the female stores the sperm from the male in a special capsule within her abdomen called a spermatheca. This reservoir allows her to continue to produce and fertilise eggs throughout her life as an adult. The female has a small catapult at the end of her body and uses this to scatter her eggs onto the ground, one by one. The eggs resemble tiny seeds and remain among the leaf litter until they are ready to hatch. Other species glue their eggs to leaves or branches or deposit them in soil or crevices.

For those unfortunate females which do not find a mate, all is not lost. Some species are capable of parthenogenesis, a process by which viable offspring can be produced without fertilisation from a male.

Stick insects undergo incomplete metamorphosis, with the young nymphs looking like miniature, wingless versions of their parents, although the hatchlings of some species resemble ants. Stick insects moult their skin several times before growing into adults, often eating their old skeletons once they have cast them off.

Defence

With their large bodies, stick insects are a prized delicacy for many predators, including birds. The hatchling nymphs are often picked off by spiders, lizards and ants on their way up a tree trunk in search of foliage. Because stick insects have no stingers and jaws that are fairly small and blunt, they must rely on other means to defend themselves against hungry predators.

The scientific name for stick insects comes from the word *Phasma* meaning ghost or apparition, referring to their ability to vanish into vegetation using their amazing camouflage. Stick insects can mimic green leaves, brown dead leaves, sticks, bark, grass and lichen. Some are so convincing that they sway gently from side to side, mimicking foliage blowing in the breeze. They can also lash out with their spiky legs, produce scary hissing noises by rubbing their wings together or shed their legs if they are attacked (much like a lizard losing its tail).

Goodie or baddie?

A few species of stick insects have made pests of themselves in the forestry industry, with populations sometimes reaching such numbers that whole stands of trees may be defoliated. However, within the realm of a backyard, stick insects generally favour native plants over a veggie garden and are virtually harmless, making them a great insect to have around.

Future Fido?

Stick insects make fantastic pets. They are easy to care for, safe to handle and won't tear up your sneakers. All you need is a good supply of fresh leaves and a tall secure enclosure.

Fascinating facts

- The titan, an Australian stick insect, holds the world record for egg-laying by stick insects – females can lay over 2000 eggs.
- Australia is home to one of the rarest insects in the world. The Lord Howe Island phasmid, found exclusively on Lord Howe Island, was thought to have been extinct since the early 1900s, when rats were accidentally introduced to the island. The slow-moving insects were easy prey for the rodents and before long the entire population was wiped out. However, in 2001, scientists discovered a small population of the insects on a nearby rocky outcrop, known as Ball's Pyramid. A breeding program has been established at Melbourne Zoo to bring them back from the brink of extinction.
- Some stick insects begin their lives in an ant nest. Foraging ants come across stick insect eggs among the leaf litter and mistake them for delicious seeds, carrying them back to their nest and storing them in a special food chamber. This is actually a good thing – deep underground, the egg will be sheltered from harsh weather and protected from fires. Upon emerging from its egg, the nymph must be careful because if the ants find it they will most certainly kill and eat it. For this reason, newly hatched stick insects often resemble ants in appearance, behaviour and smell. This clever bluff allows them

Spiny leaf insects (*Extatosoma tiaratum*) come in several colours. They can be green (to mimic a leaf), brown (to mimic a dead leaf) or, like the specimen in this photo, a very pale greenish-white (to mimic lichen). Source: K. Hiller.

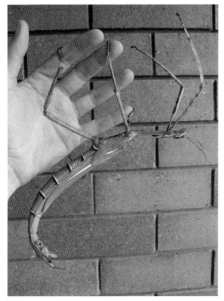

While not the longest stick insect in Australia, this female goliath stick insect (*Eurycnema goliath*) is one of the heaviest species.

enough time to leave the ant nest safely and climb into surrounding vegetation.
- The peppermint stick insect earns its nickname from a unique defensive technique, designed to deter predators such as birds. When disturbed, this turquoise-coloured insect shoots a jet of milky white fluid from a pair of special glands on its thorax. This secretion smells exactly like peppermint toothpaste and burns upon contact, especially if directed into the mouth or eyes.

Learn more

In this book:

Red-winged stick insects, p. 235
Spiny leaf insects, pp. 21, 251; ant-mimicking nymphs, p. 180

Stick insects often remain motionless during the day and move about the foliage at night to feed, when there are fewer predators such as birds. This Kirby's stick insect (*Xeroderus kirbii*) is almost invisible against the bark of a brush box.

In other books:

Rentz D (1996) *Grasshopper Country: The Abundant Orthopteroid Insects of Australia.* UNSW Press, Sydney. For beautiful photographs and detailed information on some different stick and leaf insects in Australia. ●●

Brock PD (1999) *The Amazing World of Stick and Leaf-Insects. The Amateur Entomologist*, Vol. 26. Amateur Entomologists' Society, Kent. Black and white line drawings, colour photographs and heaps of information on stick and leaf insects found throughout the world. ●●

Brock PD, Hasenpusch JW (2009) *The Complete Field Guide to Stick and Leaf Insects of Australia.* CSIRO Publishing, Melbourne. A full-colour field guide to Australian stick and leaf insects. ●●

Cleave R, Tulloch C (2015) *Phasmid: Saving the Lord Howe Island Stick Insect.* CSIRO Publishing, Melbourne. The amazing story of this stick insect's journey back from the brink of extinction, written for children and beautifully illustrated. A great read the whole family can enjoy. ●

An explanation of the above book ranking system can be found in Further Reading.

Online, type in:

Lord Howe Island phasmid – for more information on this incredibly rare stick insect and its journey back from the verge of extinction.

Stick insect camouflage – type this into an image search and prepare to test your eyesight as you squint to find the hidden insects.

Stick insect moulting – type into a video search to see great time-lapse footage of these insects shedding their outer skins.

World's largest stick insect – nicknamed 'Lady Gaga-ntuan' this enormous and elusive stick insect resides in Australia.

Termites – Order Isoptera

Summary – Termites

Order Isoptera, meaning 'equal wing'

Synonym: Blattodea (currently valid)

2864 species worldwide, 348 species in Australia

Key features

- Small, delicate and soft-bodied
- Usually pale brown, cream or yellow
- Wings present in reproductive termites, absent in soldier and worker termites
- Two pairs of equal-sized wings that are long, blade-like and membranous, folding flat over the abdomen when at rest and extending well past the end of the body

Giant northern termite worker (*Mastotermes darwiniensis*). Source: Science Image.

Don't confuse termites with:

Ants

Order Hymenoptera, see p. 174

Things in common

- Similar in size and shape (termites are sometimes referred to as 'white ants')
- Wings present in reproductive ants, absent in soldier and worker ants

Differences

- Ants have a narrow 'waist' (i.e. the point where their thorax and abdomen meet is constricted). Termites do not have this 'waist' and instead have a uniformly broad body (see Question 18, p. 48, of the identification key for an illustration of this)
- The wings of ants (when present) are unequal in size, with the fore wing larger than the hind wing. Termites have wings that are equal in size
- Ants have elbowed antennae, whereas termites have antennae that are straight

Giant bull ant (*Myrmecia gulosa*). Source: J. Green.

Description

Termites are small soft-bodied insects that are usually creamy white, yellow or brown. They are social insects, living together in large family groups in a communal nest. Within each colony you can recognise three distinct castes – groups of individuals that have a unique appearance and behaviour to help them perform certain jobs within the nest.

Workers – These sterile males and females are the smallest in stature, but the most numerous individuals within the colony. They have no wings, are usually blind and their soft pale bodies lack the protective armour seen in other castes. These poor guys get stuck with almost all of the work – collecting food, building and maintaining the nest, caring for the eggs and young, feeding the soldiers and serving the queen.

Soldiers – These guys are the bouncers of the termite world and their sole responsibility is to defend the colony. They are larger than the workers, have heavy body armour and spend most of their time standing around looking menacing.

Reproductives – Also known as alates, these males and females are the only fertile individuals in the colony. They are larger than the other castes and are equipped with two pairs of long delicate wings. Their job is to locate a new nesting place, mate and produce the thousands of eggs needed to establish a new colony.

Diet and habitat

Termites are quite small and secretive and often go unnoticed by humans. However, they are master architects and it is the elaborate nests or *mounds* they construct that usually reveal their presence. In tropical areas these mounds can rise above the ground in grasslands and open forest, like miniature mountains. Termites also occur in tree canopies, dead logs, tree stumps or timber structures, or are hidden deep within the soil.

Termites are strict vegetarians, munching their way through living and dead wood, leaves, grass, bark and leaf litter. It must be an acquired taste, but some termites devour the nutrient-rich dung of herbivores. Fungi provide an important source of protein in the diets of termites. For this reason, termites often target moist decaying wood and vegetation, which provide the ideal habitat for large populations of fungi.

Like many animals, termites sometimes struggle to break down the tough cellulose in their leafy diets. To solve this problem, they have recruited tiny symbiotic organisms called protozoa, which live in their gut and secrete enzymes to aid in digestion.

Lifecycle

At certain times of the year, thousands of winged termites leave their colony and take to the air. Males and females lock together like tandem skydivers and search for a suitable nesting site in the wood or soil. With their relocation complete, the pair chew or scrape off their wings (which only get in the way during life underground) and lay a batch of tiny eggs into a chamber. The queen cares for the first batch of eggs and passes the job over to her workers once they mature. She will continue to lay as many as 3000 eggs a day throughout her adult life (up to 17 years in some species). While male ants die shortly after mating, the king termite enjoys a long life within the nest.

Unlike other social insects such as bees and ants, termites undergo incomplete metamorphosis, emerging from their eggs as miniature versions of their parents. These nymphs gradually grow larger by moulting their skin several times.

Defence

Termites are gobbled up by echidnas, numbats, lizards and birds. But their most ferocious battles are fought against ants – sworn enemies which attack their nests and slaughter the inhabitants.

Soldier termites come in two different forms, depending on the species. Mandibulate soldiers have enormous heads armed with massive jaws – sometimes their head can be longer than the rest of their body! The strong jaws cleave off the legs and heads of attacking ants and their ridiculously large head is handy for plugging up holes in the defensive walls of their mound. Nasute soldiers have a smaller head armed with a long pointed snout. Like a fire hose, this snout showers would-be attackers with sticky glue or toxic chemicals, resulting in a hasty retreat.

Goodie or baddie?
Wood is wood to a termite, whether it be a nice tall gum tree in the bush or the timber walls of your house. They are destructive pests in our homes and their secretive behaviour means we often notice them only once the damage has been done. Luckily, only 20 or so species cause economic damage. In the wild, termites play an important role in recycling nutrients and their wood-munching behaviour produces tree hollows that are used as nesting sites for many birds and mammals.

Fascinating facts
- More than a million termites can be found in a single colony and some termite mounds stand taller than an adult man.
- During moulting, young termites can strip their bodies of the special gut fauna that helps them digest wood. But these vital supplies are easily restocked – they simply eat the poo from other termites!
- The feeding activity of termites produces the wonderful hollow tubes of wood that are fashioned into didgeridoos.
- Recently, the classification of termites has been revised and they have been moved from the Order Isoptera to an Epifamily (a special ranking that lies between a Superfamily and a Family) of cockroaches (Order Blattodea). Scientists believe that cockroaches and termites share a common ancestor and indeed, to this day, both groups share many similarities. For example, both termites and many cockroaches eat a diet consisting mainly of wood and they both use protozoa in their guts to help them break down cellulose. However, this is a recent taxonomic reshuffling: taxonomists are renowned for changing their minds and almost all books on the subject of termites still use their original classification. For these reasons, this book continues to list termites in the Order Isoptera.
- Keep Fido on a leash! Scientists have concluded that some species of subterranean termites prefer to attack wooden posts and poles that are frequently urinated on by passing dogs. It is believed that the dog's urine enriches the surrounding soil with nitrogen and protein, making it a more favourable habitat for termites than a pole that hasn't been used as a dog's restroom.

Learn more

In this book:

Giant northern termites, p. 175

In other books:

Hadlington P, Marsden C (1998) *Termites and Borers: A Homeowner's Guide to Detection and Control.* UNSW Press, Sydney.

This giant northern termite soldier (*Mastotermes darwiniensis*) is more heavily armoured than the surrounding soft-bodied workers. Being a mandibulate soldier, it uses its large powerful jaws to defend the colony. Source: Science Image.

The grotesquely swollen body of the queen termite (*Nasutitermes exitiosus*) dwarfs those of her workers and soldiers. At the top right of the photo are nasute solider termites that use their long pointed snouts to spray enemies with sticky glue or toxic chemicals. Source: Forestry and Forest Products.

A handy, easy to read guide on recognising the presence of termites in and around your home, methods for preventing and controlling infestations and frequently asked questions about termites. ●

Hadlington P, Staunton I (2008) *Australian Termites*. 3rd edn. UNSW Press, Sydney.

Easy to read information on the biology of termites, including their diets, nest construction and lifecycles. It features identification keys which can be used (with the aid of a microscope/magnifier) to identify various Families and Species of Australian termites, complete with detailed diagrams and distribution maps. Chapters on recognising the presence of termites, methods for preventing and controlling infestations and detailed case studies makes this an excellent resource for home-owners concerned about termite control. ●●

An explanation of the above book ranking system can be found in Further Reading.

Online, type in:

Nasute soldier termite – see images and learn more about this special form of soldier termite which sprays its foes with toxic chemicals.

Soldier termites – see images and learn about the special modifications these termites have to help protect their colony.

Termite mound images – for photos of different types of termite mounds, including some that are taller than an adult male.

Termite queen – type into an images search to see photographs of these enormous egg-laying machines.

Thrips – Order Thysanoptera

Summary – Thrips

Order Thysanoptera, meaning 'fringed wing'

5749 species worldwide, 738 species in Australia

Key features

- Tiny elongate bodies, often black
- Two pairs of long narrow wings with a fringe of tiny hairs along their margins; wings sometimes absent

Cycad thrips (*Cycadothrips chadwicki*). Source: J. Dorey.

Don't confuse thrips with:

Aphids

Order Hemiptera, Suborder Sternorrhyncha, see p. 278

Things in common

- Tiny, sometimes black
- Wings sometimes absent
- Like thrips, aphids are commonly found on plants

Green peach aphid (*Myzus persicae*). Source: J. Dorey.

Differences

- Aphids have shorter stouter bodies than thrips
- Aphids have two tubes at the end of their bodies, which are absent in thrips
- When present, aphids' wings are held roof-like over the body, whereas thrips fold them flat
- The wings of aphids lack a fringe of hairs

Ants

Order Hymenoptera, see p. 174

Things in common

- Often similar in size, shape and colour

Differences

- Ants have a narrow 'waist' (i.e. the point where their thorax and abdomen meet is constricted). Thrips do not have this 'waist' and instead have a uniformly broad body
- Ants have elbowed antennae, whereas thrips have antennae that are straight

Sugar ant (*Camponotus* sp.).

Description

Thrips are small to minute insects, ranging in size from 0.5 mm to 15 mm long. Their tiny size means they usually go unnoticed, despite being abundant across most of the world.

Thrips have bodies that are small, slender and slightly flattened and some have a tapered tube at the end of their abdomens. They are usually brown, black or yellow, although the immature stages may be pink, orange or reddish. Some adults have two pairs of fringed wings – narrow wings surrounded by a margin of long hairs. At rest, these wings are held parallel to their bodies or crossed over their abdomens.

Thrips have remarkable feet (tarsi). Instead of the usual claws that adorn the ends of most insect legs, thrips have a tiny air-bag called an arolium which can be inflated on command to help them adhere to slippery surfaces.

Diet and habitat

Thrips usually go unnoticed by humans, happily hiding away in flowers, on leaves and bark and in leaf litter. However, at certain times of the year populations explode and thrips amass in the air, on glass windows and sometimes on washing on our clotheslines.

Thrips have a tiny piercing mouth, which they inject into pollen grains, the cells of leaves, floral tissue and fungi, sucking out the contents. Some are predators, feasting on small insects and mites. Others form tiny galls, which resemble small 'pimples' on the surface of leaves (see p. 150).

Lifecycle

Thrips have a very puzzling lifecycle that lies somewhere between incomplete and complete metamorphosis. They scatter their eggs on the surface of plants or insert them directly into plant tissue. From these eggs hatch tiny larvae which closely resemble their parents, except they lack wings and are pale and often translucent. They moult their skin a couple of times before becoming a pre-pupa. This is a resting stage and during this time the tissue within the thrips undergoes a major breakdown and reformation, similar to that of a caterpillar turning into a butterfly. For this reason we refer to the young as 'larvae' rather than 'nymphs'. They emerge from this stage as an adult, which, depending on the species, may or may not have wings.

Adult thrips are capable of parthenogenesis, which means they can produce fertile eggs without needing to mate.

Defence

Due to their tiny size, thrips sit at the bottom of the food chain and are picked off by a variety of other animals. Tiny flies, true bugs, mites, lacewing larvae and even other species of thrips are all known predators. Small wasps scoop them up and store them in their nests as food for their larvae. Other wasps parasitise thrips – laying eggs directly into their bodies which hatch into tiny maggots, eventually killing and consuming them.

Some thrips try to ward off predators by making themselves look bigger, lifting the tip of their abdomen up over their head in a threatening gesture. Many thrips larvae take a more controversial approach. When confronted, they use their long flexible bodies to smear their attacker with liquid faeces! The sensory organs of the hungry predator are fouled up with this horrible mess, forcing it to abandon the attack in favour of a much-needed grooming session.

Goodie or baddie?

Thrips can wreak havoc in our vegetable gardens, glasshouses and crops. Their feeding and egg-laying can inflict severe damage, causing some plants to drop their leaves. Their method of feeding can also transmit plant viruses, in much the same way that mosquitoes spread disease. The tomato spotted wilt virus is one such disease and is a major risk to fruit crops in Australia.

On the other hand, thrips frequently visit flowers and can play an important role in the pollination of some plants. Others are predators of several species of mites that are known pests in agriculture.

Fascinating facts
- There is no such thing as a 'thrip'. They are always referred to as 'thrips', even if there is only a single individual.
- Thrips are left-handed. Well, to be more correct, they are left-mouthed. They have asymmetrical mouthparts, with the right mandible (jaw) reduced or completely absent. So, it is the left mandible that forms the narrow tube which pierces the food.
- Thrips can really get around. Even the wingless forms can drift from place to place on air currents and they have been known to blow from Australia across the Tasman Sea, all the way to New Zealand.
- The grain thrips is a rather pesky creature. In Europe, they form huge swarms in late summer. These clouds of tiny insects can invade buildings through cracks and crevices and their movement can trigger fire alarms. People exposed to swarms can be bitten, causing itchy, often painful, rashes.
- More on the grain thrips. A 59-year-old farmer repeatedly visited a dermatology clinic in Italy, complaining of consistent itching on her scalp and the sensation of insects crawling on her head. However, no evidence of insects could be found and she was diagnosed with Ekbom's syndrome, a psychological condition in which the victim has a strong delusion that they are infested with some sort of bug or parasite. Interestingly, a subsequent visit to the clinic revealed several grain thrips were indeed crawling over her scalp and irritating the skin. It turns out the patient's proximity to wheat farms brought her into direct contact with swarms of grain thrips – after a month's stay away from the farm, all her symptoms disappeared.
- Australian cycads in the Genus *Macrozamia* rely exclusively on a certain kind of thrips for pollination. These ancient plants are made up of male and female individuals, each of which produces cones for reproduction. Male cones are laden with pollen, providing food for the thrips. Male and female cones release low levels of special air-borne chemicals that attract thrips. However, between the hours of 11am and 3pm, these emissions soar to toxic levels. At the same time, the cones

Giant thrips (*Idolthrips spectrum*) can be found in the hundreds on dead eucalypt leaves, making them one of the most common insects in Australia. Source: T. Daley.

begin to self-heat, increasing their temperature up to 20°C hotter than their surrounding environment. These factors work together to drive thrips from cones – most importantly, pollen-covered thrips from male cones. Later in the day the cones begin to cool and the chemical emissions drop, once again providing an inviting habitat for thrips. Female cones do not provide food; however, their chemical smell is very similar to that of male cones. Pollen-covered thrips mistakenly enter female cones in search of food, facilitating pollination. This amazing relationship is known as 'push–pull' pollination – 'pushing' pollen-covered thrips from male cones and later luring or 'pulling' them back into cones to drive pollination. It is this fragile relationship between plant and pollinator that makes cycads one of the most threatened groups of plants worldwide.

Learn more

In other books:

See Further Reading

Online, type in:

Thrips garden pests – lots of fact sheets from gardeners on how to recognise thrips damage and control these pests in your garden.

Thrips' wings – type into an images search to see lots of close-up images of thrips, showing their amazing fringed wings.

True bugs – Order Hemiptera

Summary – True bugs

Order Hemiptera, meaning 'half wing'

100 428 species worldwide, around 6000 species in Australia

Key features

- Highly variable in shape and size
- Mouth always modified into a piercing and sucking tube
- No cerci (finger-like appendages) at the end of the abdomen

Aphids, mealybugs, whiteflies, scale insects and their allies, Suborder Sternorrhyncha

Synonym: Homoptera

Key features

- Small and soft-bodied
- Two pairs of membranous wings, held roof-like over abdomen at rest

INSECT ORDERS 275

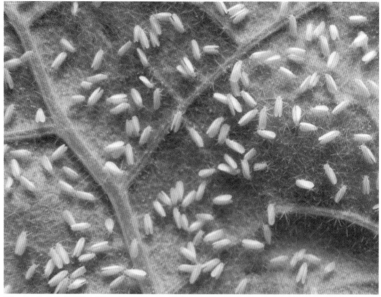
Silverleaf whitefly (*Bemisia tabaci*). Source: D. Papacek.

- Wings often absent, or only fore wings present in some groups
- Antennae usually long and thin
- Often construct shelters (e.g. wax scales, fluffy filaments or hardened shells) to protect their delicate bodies

Cicadas, leafhoppers, planthoppers and their allies, Suborder Auchenorrhyncha

Leafhopper (*Rosopaella* cf. *crofta*). Source: T. Daley.

Synonym: Homoptera

Key features

- Two pairs of wings, held roof-like over the abdomen at rest
- Bristle-like antennae
- Large compound eyes
- Spines on their back legs, often used for jumping

Stink bugs, water bugs, plant bugs and their allies, Suborder Heteroptera

Synonym: Geocorisae

Key features

- Two pairs of wings, folded tight and flat over abdomen at rest
- Fore wings entirely or partially hardened into a leathery shell (hemelytra) which covers the hind wings
- Antennae usually long and thin
- Large compound eyes

The information pages for each Suborder have information on other insect Orders you may have them confused with.

Shield bug (cf. *Hiarchas angularis*). Source: K. Ellingsen.

Description

While many of us use the term 'bug' to cover any kind of creepy-crawly that crosses our path, a 'true bug' belongs to the Order Hemiptera. This is a very large and diverse Order that includes, among many others, aphids, mealybugs, scale insects, whiteflies, cicadas, leafhoppers, froghoppers, stink bugs, assassin bugs, water bugs, shield bugs, jewel bugs, plant bugs and crusader bugs. They come in a wide variety of shapes, sizes and colours. Some are so small they are difficult to see with the naked eye, others are whoppers up to 11 cm long.

Planthoppers, with their well-developed wings and strong jumping legs, are highly mobile, whereas female scale insects are wingless legless blobs that adhere to the surface of plants as adults. Cicadas have clear wings with a series of complex veins, while jewel bugs keep their relatively simple wings tucked underneath a brightly coloured shield on their backs. There are even true bugs that have managed to conquer aquatic habitats – our waterways are teaming with water scorpions, backswimmers and giant water bugs. The one feature that unites these seemingly unrelated creatures is a unique modification to their mouthparts.

The fact is that true bugs suck – literally! The main distinguishing feature of this Order is the highly modified, piercing–sucking mouthparts. While most insects chew their food with strong jaws, true bugs possess a pair of sharp, barbed needle-like stylets that form two pipes within a piercing and sucking tube bundled together in a sheath. This structure is known as a rostrum and is used to extract fluid from the living tissue of plants and animals. The stylets pierce the surface and probe around for a suitable place to feed. One channel within the rostrum pumps saliva into the site, flooding it with enzymes to help break down the tissue. A second tube draws up the liquefied food, using a strong sucking pump.

The piercing–sucking mouthparts are the one unifying feature of true bugs, for all their other features vary greatly – the structure of their wings and how they hold them at rest, the way they use their legs, where they live, what they eat, and their general shape, size and colour.

The highly diverse Order Hemiptera is divided into separate Suborders and each is dealt with separately. Suborder Sternorrhyncha (p. 278) includes the small, usually soft-bodied bugs such as aphids, mealybugs, whiteflies and scale insects. Cicadas, leafhoppers and planthoppers have strong armoured bodies and wings with a complex network of veins, and belong to Suborder Auchenorrhyncha (p. 285). Heteroptera (p. 292) is the largest and most diverse Suborder; it includes stink bugs, water bugs and plant bugs. Members of this group have parts of their fore wings hardened into a leathery protective cover. A fourth Suborder, Coleorrhyncha (not covered in this book), consists of a single Family of insects known as moss bugs that have only recently been categorised into their own Suborder.

Aphids, mealybugs, whiteflies, scale insects and their allies

Summary – Aphids, mealybugs, whiteflies, scale insects and their allies

Glasshouse whitefly (*Trialeurodes vaporariorum*). Source: T. Daley.

Order Hemiptera, Suborder Sternorrhyncha

Around 16 000 species worldwide, around 1413 species in Australia

Key features
- Tiny, usually soft-bodied
- Two pairs of membranous wings, held roof-like over abdomen
- Wings may be absent, or only the fore wings present in some groups
- Antennae usually long and thin

Don't confuse aphids, mealybugs, whiteflies, scale insects and their allies with:

Booklice

Order Psocoptera, see p. 194

Things in common
- Tiny

Booklouse (*Propsocus pulchripennis*). Source: T. Daley.

- Two pairs of membranous wings, held roof-like over abdomen, wings may be absent
- Booklice resemble aphids and whiteflies and are hard to tell apart with the naked eye

Differences
- Booklice have much longer thinner antennae than aphids and whiteflies and they have bulging compound eyes
- Aphids have two (often black) tubes at the end of their bodies, which are absent in booklice
- Booklice have slightly flattened bodies, whereas aphids are much rounder in appearance

Flies

Order Diptera, see p. 225

Things in common
- Some small flies are similar in shape and size to aphids and whiteflies

Differences

- Hard to tell apart with the naked eye, but flies have only one pair of wings, whereas aphids and whiteflies usually have two pairs
- Flies have shorter antennae and usually much larger eyes than aphids and whiteflies

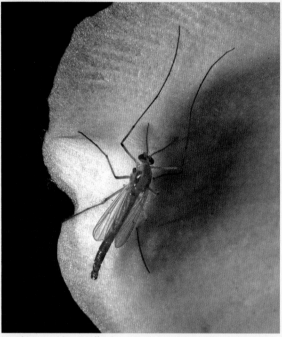

Non-biting midge (*Kiefferulus* sp.). Source: James Niland, Attribution 2.0 Generic.

Description

The Suborder Sternorrhyncha comprises a variety of small, soft-bodied insects, including (among others) aphids, mealybugs, whiteflies and scale insects. Members of this Suborder are generally called 'sternorrhynchans', a word that is quite laborious to read (and spell)! For this reason, sternorrhynchans are herein referred to as 'soft bugs' – an appropriate nickname, given their bodies are not as heavily armoured as other types of true bugs. While soft bugs vary greatly in appearance they can all be characterised by their small size and piercing–sucking mouthparts.

Aphids are plump and usually green, pink or yellow. They have long thread-like antennae slicked back over their bodies. Many of us encounter these creatures as they cluster together on the tips of our rose bushes and other ornamental plants. Whiteflies (despite their name they are true bugs, not flies) are also frequent visitors to our gardens and clouds of them can be seen rising into the air as you brush past parsley plants. They are tiny delicate insects with bodies and wings covered in a waxy white powder. Psyllids, also known as jumping plant lice, have strong hind legs that enable them to leap into the air when disturbed. Mealybugs adorn their bodies with long white threads and can be found crawling on your citrus plants and the underside of palm fronds.

Some members of this Suborder do not look like insects at all. Scale insects cover their soft bodies in protective coverings, made from substances they secrete from their bodies. These may consist of hard scaly plates, fluffy waxy filaments or, in the case of lerp insects, delicate lace-like coverings. Some of these insects live within galls (see p. 150) – a tumour-like deformity on the leaves and stems of plants. These weird growths are caused by the feeding action of some aphids, scale insects, psyllids and mealybugs, and it is within these structures that the tiny insects shelter and feed.

Diet and habitat
All soft bugs are strict vegetarians, using their piercing–sucking mouthparts to feed on plants. They target phloem, the living tissue within plants responsible for the transportation of nutrients. It is a rich source of energy and amino acids, but a lot of excess water and sugar is unavoidably ingested as the bug sucks up its meal. For this reason, many soft bugs have a specialised digestive system. A filter removes surplus water and sugar, bypasses the main digestive system (where the other nutrients are processed) and transfers it straight to the hind gut, where it can be rapidly expelled. This sugary waste product is known as honeydew. Honeydew may be squirted from the body to eliminate it from the system, kept and used to manufacture a protective scale-like covering, or excreted from the abdomen in tiny droplets, which are collected and consumed by other insects such as ants.

Soft bugs most commonly target flowering plant species, including native trees such as eucalypts and wattles, as well as many ornamentals, fruit trees, herbs and vegetables. Many species of aphids and scale insects have been accidentally introduced to Australia. Due to their tiny size and cryptic form, they often go unnoticed on agricultural plants or on ornamentals brought over from other countries. Some have been here a long time.

Lifecycle
Soft bugs undergo incomplete metamorphosis, a lifecycle that traditionally has three stages – egg, nymph (the juvenile stage) and adult. In many insect groups (e.g. cockroaches and grasshoppers) this type of lifestyle is relatively simple, with the younger stages looking like miniature versions of the adult, albeit without wings. However, within the soft bugs things are a bit more complicated, and males, females and nymphs within the same species can vary greatly in appearance.

Lerp insects have a more straightforward lifecycle than that of other soft bugs. The males and females look the same and both have wings for flying. The female uses a sharp blade at the end of her abdomen to cut slits into the tissue of plants, where she deposits her eggs. The wingless nymphs resemble their parents and remain sheltered under elaborate waxy or lacy shields until they grow into adults.

Whiteflies are also winged in both sexes. From the female's eggs hatch crawlers – tiny, highly mobile nymphs which disperse to new feeding sites. So small are these tiny hatchlings, that they are often carried by the wind to new locations. Once they have settled into their new home, they shed their skin and become sedentary insects, sheltering their bodies in waxy or scaly coverings and settling down to feed and grow. To transform into winged adults, the nymphs enter into a lengthy motionless stage, in which significant changes to their body (e.g. the formation of wings) take place. This stationary insect is referred to as a 'pupa', but it differs significantly from the similarly named stage seen in insects that undergo complete metamorphosis, such as butterflies.

The appearance of adult scale insects varies greatly between the sexes. Females are not very attractive to look at, with their bloated wingless and sometimes legless bodies. These immobile insects anchor themselves to a plant with their piercing mouthparts, often covering their soft bodies with waxy filaments or a hard shell-like cover to protect themselves. Males, on the other hand, have wings and legs and are quite mobile, actively seeking out females to mate with. But the clock is ticking for these frisky males – they lack mouthparts and cannot feed, so are fairly short-lived. The females shelter their eggs underneath their bodies, until they hatch into tiny mobile crawlers. Once settled, young scale insects grow much like whiteflies, with motionless nymphs and a 'pupal' stage during which males grow their wings.

The nymphs of mealybugs remain active throughout their lifecycle, using their long legs to move between feeding sites. Their bodies are covered in white waxy filaments. The females resemble the nymphs in appearance, whereas males emerge from a pupa with tiny wings.

Aphids have possibly the most complicated lifecycle of all the soft bugs, although the males, females and nymphs all look similar. At certain times of the year, winged males and females mate and disperse. Females lay eggs, or sometimes give birth to live young. These nymphs eventually grow into wingless females, known as stem mothers. These females produce hundreds of daughters without needing to mate with a male, a process known as parthenogenesis. This goes on for several generations, with only female insects produced, until the weather conditions are favourable for the establishment of new colonies. At this point, females start producing nymphs that grow into winged males and females. After mating, the females disperse to a new place to start the lifecycle all over again.

Defence
Due to their tiny size, soft bugs are often overlooked by larger animals such as birds and lizards. But for hungry insect predators, they make the perfect size snack. Ladybird beetles, their young and the maggot-like larvae of hover flies are ruthless hunters of aphids, eating hundreds of individuals throughout their life. Mealybugs also fall prey to ladybird beetles, as well as to the voracious larvae of lacewings. Tiny wasps inject their eggs into scale insects and their hatchling larvae devour the host from the inside out.

Soft bugs lack the strong jaws and painful stingers used by other insects to fight back against their enemies, and therefore enlist the help of those better equipped for the job. Ants go crazy for the honeydew produced by soft bugs and will hang around insects such as aphids and scale insects, greedily sucking up the sweet fluid as it exits their bodies. So reliant are the ants on their sugar fix that they fiercely guard the bugs, chasing away any potential enemies. This is a great example of mutualism, a relationship between two organisms that is beneficial to both. The ants are provided with a continual source of nourishment in return for their service as bodyguards.

Goodie or baddie?
With their piercing–sucking mouthparts, aphids, mealybugs, whiteflies and scale insects are notable pests of vegetable crops and fruit trees, including potatoes, citrus and peaches. Their sap-sucking behaviour weakens plants, causing wilting, deformities and stunted growth. These insects may transmit plant diseases, with viruses drawn from infected plants and spread to healthy ones as the bugs move between feeding sites. Honeydew can encourage the growth of sooty moulds on leaves and attracts ants, which drive away beneficial predatory insects.

Fascinating facts
- Remember manna, described in the Bible as fine wafer-like flakes that fell from the sky, providing food for the Israelites during their pilgrimage in the desert? It is likely that these sweet morsels were waxy honeydew coverings produced by a desert-dwelling scale insect.
- The crawlers of some scale insects and mealybugs hitch a ride with their winged older brothers, clinging to their bodies as they fly off in search of females. This enables them to disperse considerable distances, without the 'legwork'.

This adult female mealybug (*Monophlebulus* sp.) slowly wanders around eucalyptus, acacia and other trees, where she feeds on plant sap.

- Cochineal dye produces orange and red colours used to tint foods, drinks, cosmetics and fabrics. This dye comes from a type of scale insect that feeds on a species of cactus in South America. The bright pigment is carminic acid – a toxin produced within the scale insect's body to deter predators. The scale insects are painstakingly removed from the cacti by hand, plunged into boiling water, dried, then finely ground and mixed with calcium salts to produce the dye.
- Pea aphids come in a variety of colours, including pinkish-red. This is due to organic pigments known as carotenoids, the same substances that give certain fruits and vegetables their bright colours. Aphids are unable to produce these pigments themselves and cannot gain them from the food that they eat. Instead, they steal them from fungi. Fungi have a certain group of genes that enable them to make carotenoids. When scientists examined the DNA of pea aphids, they found this very same set of genes. Somehow, over time, the genes of the fungus became incorporated into the bodies of pea aphids, either through the fungus infecting the aphids or perhaps as a result of some sort of mutually beneficial arrangement between the two organisms. Being red has its advantages – studies have shown that certain parasitic wasps prefer to lay their eggs on green aphids rather than red ones.

Attached to the body of this female cottony cushion scale (*Icerya purchasi*) is a fluted white egg sac. This egg sac may be twice the length of the insect's body and contain as many as 1000 eggs.

Learn more

In this book:

Aphids, p. 106
Armoured scale insects, pp. 151–152
Black citrus aphids, p. 245
Cottony urbicola scale insects, p. 192
Galls, p. 150
Green peach aphids, p. 271
Lerp insects, p. 151
Mealybugs, pp. 152–153
Onion aphids, p. 194
Silverleaf whiteflies, p. 275
Soft scale insects, p. 152

In other books:

Blanche R (2012) *Life in a Gall: The Biology and Ecology of Insects that Live in Plant Galls*. CSIRO Publishing, Melbourne. Interesting information and photographs on different gall-forming insects.
●●

An explanation of the above book ranking system can be found in Further Reading.

Online, type in:

Aphid mummies – learn about the tiny parasitoid wasp that is being used to control pest populations of aphids in agriculture. After hatching from a tiny egg, the wasp larva kills and consumes its aphid host then uses the mummified carcass as a 'cocoon' where it completes its transformation into an adult wasp.

Cochineal insects – learn more about how these insects are used to produce red food colouring.

Control scale insects – heaps of fact sheets on how to control scale insects on both indoor and outdoor plants, including many pesticide-free solutions.

Lerp insects – type into an images search to see the diverse shapes, sizes and colours of lerps.

Cicadas, leafhoppers, planthoppers and their allies

Summary – Cicadas, leafhoppers, planthoppers and their allies

Order Hemiptera, Suborder Auchenorrhyncha

Around 42 000 species worldwide, around 1532 species in Australia

Key features

- Two pairs of wings, often transparent, held roof-like over the abdomen at rest
- Fore wing larger than hind wing
- Bristle-like antennae
- Large compound eyes

Bottle cicada (*Glaucopsaltria viridis*). Source: K. Ebert.

Don't confuse cicadas, leafhoppers, planthoppers and their allies with:

Bees and wasps

Order Hymenoptera, see p. 162

Things in common

- Often similar in shape and size
- Two pairs of wings, fore wing larger than hind wing
- Large compound eyes

Differences

- Bees and wasps have a narrow 'waist' where their thorax and abdomen meet, whereas cicadas, leafhoppers and planthoppers usually have a uniformly broad body
- Bees and wasps often have much longer antennae than cicadas, leafhoppers and planthoppers
- Bees and wasps tend to be very active, whereas cicadas, leafhoppers and planthoppers often perch on the branches and trunk of trees

Honey bee (*Apis mellifera*). Source: D. Papacek.

Flies

Order Diptera, see p. 225

Things in common

- Often similar in shape and size
- Large compound eyes
- Bristle-like antennae

Differences

- Flies have only one pair of wings, whereas cicadas, leafhoppers and planthoppers have two pairs
- Flies do not hold their wings roof-like over their abdomens while at rest, but instead fold them flat over the abdomen or alongside the body
- Flies tend to be very active, whereas cicadas, leafhoppers and planthoppers often perch on the branches and trunk of trees

Snail parasitic blowfly (*Amenia* sp.).

Description

The Suborder Auchenorrhyncha includes (among others) cicadas, spittle bugs, leafhoppers and planthoppers. These insects have a characteristic way of holding their two pairs of membranous wings while at rest – like a little roof or an upside-down V over their abdomen. They share the piercing–sucking rostrum (needle-like mouth) seen in other members of the Order Hemiptera and are generally larger and more heavily armoured than their tiny, soft-bodied aphid cousins.

Members of the Suborder Auchenorrhyncha are generally called 'auchenorrhynchans' but to make things more simple they are referred to here as 'cicadas and hoppers' – in honour of some of the high-profile members of this Suborder.

Cicadas are the largest and most easily recognised members of this group. Their wings are generally clear and covered in a network of large veins, although some species, like the bladder cicada, have opaque green wings that resemble leaves. Cicadas have large bulging compound eyes and their antennae are reduced to a pair of tiny hair-like bristles, which are difficult to see with the naked eye.

Leafhoppers, treehoppers, froghoppers and planthoppers (here grouped together as 'hoppers') look like miniature versions of cicadas but their fore wings are usually thick, leathery and opaque and cover their delicate hind wings. As their name suggests, these insects are gifted jumpers and use their spiny hind legs to help them leap into the air.

Diet and habitat

Cicadas and hoppers are frequent visitors to our gardens, but some can be quite cryptic and hard to find. Cicadas are often heard but not seen, as they shelter high among the branches of eucalypts and other trees. Many treehoppers and leafhoppers sit motionless on the stems of plants, resembling thorns or buds at first glance. Others such as planthoppers are more noticeable, aggregating in large masses on the stems of plants.

Cicadas and hoppers are quite fussy eaters, with all members following a strict vegetarian, liquid-only diet. They plunge their rostrum into the living tissue of plants, targeting the phloem and xylem, which are responsible for the transportation of water and nutrients. Cicadas and hoppers have a specialised digestive system designed to filter excess sugar and water (a by-product of their sap-based diet) straight to the end of the body where the excess water and sugar can be rapidly flushed out. This waste product, known as honeydew, may attract insects such as ants, which greedily feed upon the sugary secretions as they are expelled from the abdomen.

Most cicadas and hoppers feed on trees and shrubs, with many species targeting the stems and shoots of native plants such as eucalypts, wattles and casuarinas. However, some groups feed underground on roots, in clumps of grasses, on the trunks and bark of trees, or on ornamental and agricultural plants.

Lifecycle

Cicadas are well known for their singing abilities and males gather together on tree trunks to attract and serenade nearby females. To attract the appropriate partner, each species of cicada has a unique call, much like birds. Sound is produced through special organs known as tymbals – a pair of thin membranes located on either side of the abdomen, hooked up to a network of internal muscles. Rapid contractions of these muscles cause the tymbals to click in and out very fast (like one of those old-fashioned metal clicker toys), with some species clicking 1050 times in a single second. Large air-filled sacs within the abdomen help to amplify the sounds, broadcasting these melodies to the females in the surrounding area. Cicadas are also equipped with sophisticated sound-receiving organs known as tympana, so that the vocal efforts of the males do not fall upon deaf 'ears'. Located just below the tymbals in both male and female cicadas, these large oval-shaped membranes are covered in tiny sensitive hairs and concealed beneath protective plates.

Other members of this Suborder use sound to attract potential mates, albeit in a more subtle manner than the raucous songs of cicadas. Many leafhoppers and planthoppers produce non-audible calls by perching on the stems of plants and vibrating or clicking parts of their bodies. These secret songs travel through the stem and are picked up by other individuals resting on the plant. We need special equipment to amplify these calls so we can hear them.

After mating, the females use a sharp blade at the end of their abdomen known as an ovipositor to cut slits into plant tissue, where they deposit their eggs. Cicadas and hoppers have a fairly simple lifecycle known as incomplete metamorphosis, where the younger stages (nymphs) develop wings in small external sheaths on their backs. Their tough outer skin must be shed several times to accommodate their growing bodies and evidence of this phenomenon may be left behind for us to discover.

Cicadas shells are often found clinging to vertical surfaces, such as the trunks of trees and fence palings. These 'shells' are the final exoskeleton from the cicada nymph, a strange hump-backed creature that lives underground, feasting on the roots of trees. When the cicada nymph reaches a certain size it tunnels its way out of the dirt, often at night when there are fewer predators such as birds. The nymph then climbs up a tree trunk and clings tightly to it, using sharp claws on its fore legs for grip. Over several hours the cicada splits open the back of its exoskeleton and extracts its body, legs and newly formed wings. Initially small and crumpled, these wings expand and harden as the cicada pumps blood through a network of tiny veins.

The nymphs of planthoppers and leafhoppers can be found in similar habitats to their parents. Both generations often feed side by side on the same plant.

Defence
Cicadas are fairly large and can be spotted as they fly between trees in search of feeding sites, making them an easy target for aerial predators such as birds. Cicadas are not known for their endurance and can generally only fly short distances. When confronted by a hungry predator they switch to an evasive flight pattern, zigging and zagging through the air, rapidly changing direction in an attempt to outmanoeuvre their enemies. They also use their tymbals to produce a loud alarm cry to startle would-be predators.

Many hoppers hide from their enemies. Treehoppers often have elaborate protrusions poking out from their bodies. These 'horns' help the insects, which are generally brown, to disguise themselves as thorns as they perch on the stems of plants. Some leafhoppers have extremely flat, greyish brown bodies, which are great for blending into bark. Others hide, quickly darting behind the stems of plants if danger approaches. However, if all else fails these insects do exactly as their names suggest and use their strong hind legs to quickly hop out of harm's way.

Several species of planthoppers attract ants, which feed upon the honeydew secreted from their bodies as waste. This bevy of bodyguards protects the bugs in exchange for the sugary meals.

Froghopper nymphs construct unusual shelters to protect their tiny bodies. One type, appropriately called spittle bugs, cover themselves in a frothy fortress of sticky bubbles made from a combination of waste products mixed with air. These white blobs, which drip liquid, are often very obvious, numerous and annoying to the owners of cars parked beneath them. Others species construct hard scaly tubes fastened to the stems of plants and shelter within their walls.

The elaborate projection on this horned treehopper (*Eutryonia monstrifer*) helps it to mimic a thorn as it rests on a plant stem. Source: J. Dorey.

This unusual leafhopper (*Stenocotis depressa*) has an extremely flattened body so that it does not cast a shadow (which may reveal its location to a predator) as it rests on the trunks of eucalypt trees.

Goodie or baddie?

Cicadas and hoppers may damage plants through their piercing–sucking method of feeding. As with their aphid cousins, some leafhoppers cause wilting, deformities and stunted growth in vegetable and fruit crops and are also capable of transmitting plant viruses. However many of them, such as cicadas and planthoppers, restrict their feeding to native plants such as eucalypts and wattles and so lack the agriculturally destructive potential of their aphid and scale insect cousins.

Fascinating facts

- Have you ever wandered underneath a gum tree on a fine sunny day, only to have droplets of water rain down on you? Whatever you do, don't stick out your tongue to wet your whistle! This is no ordinary rain – it is cicada rain. Basically, you are being peed on by

Passionvine hoppers (*Scolypopa australis*) feed on a variety of plants and can be a pest of vine crops such as kiwifruit and passionfruit. Source: K. Ellingsen.

hundreds of cicadas feeding in the tree, expelling the excess honeydew from their bodies as they feast on the plant sap.

- In the spring, fully grown cicada nymphs leave their underground tunnels and burrow upwards, where they pause just below the surface. There they wait until the right weather conditions, usually a warm wet day, trigger their emergence. If the weather is not favourable, they burrow back underground and wait patiently until the next spring to try again.
- Froghoppers, as their name suggests, are renowned for their jumping ability. It takes a fraction of a second to launch their bodies up to 70 cm into the air. Admittedly, this is little more than knee height on your average human, but it adds up to a whopping 115 times the body length of an adult froghopper. To put it into context, it would be like us jumping over a 70-storey building.
- The nymphs of treehoppers have an 'anal whip' – a long expandable tube that extends from their anus. Exactly why they have this unusual appendage is a topic for some debate. Some think it is used for defence – treehoppers readily thrash it about from side to side if they are handled. Or it may be a handy extension for expelling waste – reaching out to deposit drops of honeydew away from the body, so that the nymph does not get covered with sticky unpleasantness.
- Although we are unable to hear it, the plants in our gardens are literally pulsing with a symphony of secret music. As mentioned earlier, some leafhoppers and planthoppers transmit vibrations through the stems of plants. A microphone hooked up to a plant stem covered in leafhoppers allows scientists to eavesdrop on their seductive songs.

Learn more

In this book:

Brown bunyip cicadas, p. 242
Cicadas, p. 107
Dodd's bunyip cicadas, p. 107

Eucalypt planthoppers, p. 125
Lantana treehoppers, p. 142
Leafhoppers, p. 275
Razor grinder cicadas, p. 227
Spittle bugs, p. 154

In other books:

Moulds MS (1990) *Australian Cicadas.* UNSW Press, Sydney.
A detailed guide to the different species of cicadas in Australia. For each species listed there is a distribution map, notes on their habitat, songs and life history and a colour photograph. ●●
An explanation of the above book ranking system can be found in Further Reading.

Online, type in:

Cicada lifecycle – type this into a video search to see heaps of amazing movies (including a great documentary from Sir David Attenborough) showing cicada nymphs emerging from the ground and moulting their skins to reveal their wings.
Lantern fly – type this into an image search to see these amazing true bugs with weird projections on their heads (including the aptly named peanut-headed lantern fly).
Membracidae – this is the family to which treehoppers or thorn bugs belong. Type it into an image search to see these masters of mimicry and camouflage at their most weird and wonderful.
Periodical cicadas – learn more about this North American species that emerges *en masse* every 17 years.

Stink bugs, water bugs, plant bugs and their allies

Summary – Stink bugs, water bugs, plant bugs and their allies

Order Hemiptera, Suborder Heteroptera

Over 42 300 species worldwide, 2518 species in Australia

Key features

- Two pairs of wings that sit tight and flat over abdomen at rest
- Fore wings entirely or partially hardened into a leathery shell (hemelytra) which covers the hind wings
- Antennae usually long and thin
- Large compound eyes
- No cerci (finger-like appendages) at the end of abdomen

Spined predatory shield bug (*Oechalia schellenbergii*) with egg mass. Source: D. Papacek.

Don't confuse stink bugs, water bugs, plant bugs and their allies with:

Beetles

Order Coleoptera, see p. 186

Things in common
- Fore wings hardened into a shell-like cover that sits tight and flat over the abdomen at rest
- Often similar in shape and size – it can be quite difficult to tell bugs and beetles apart
- No cerci at the end of abdomen

Differences
- In beetles, the fore wings form a shell that meets in a straight line. In bugs, the wings overlap forming a cross, or no line is visible at all (see Question 6 in identification key, p. 35)

Fiddler beetle (*Eupoecila australasiae*).

Description

Members of the Suborder Heteroptera are sometimes referred to as 'true bugs', however this name is more commonly used to describe **all** insects belonging to the Order Hemiptera. To avoid confusion, insects belonging to the Suborder Heteroptera are herein dubbed 'typical bugs', an appropriate name given that this Suborder contains the large, more characteristic 'bugs' that most of us are familiar with.

Typical bugs have a unique modification to the design of their wings. Like beetles, these insects have hardened fore wings that form protective coverings, which fold tight and flat over their delicate hind wings. However, unlike the rigid elytra (wing covers) of beetles, typical bugs have fore wings that are leathery and flexible. Known as hemelytra, they allow both pairs of wings to be used in flight, while still providing protection for the membranous hind wings. In most typical bugs only the base of the fore wing is hardened, with the membranous tips clearly visible as they rest, overlapping, across the end of the abdomen. As usual, we find some exceptions. In many water bugs, for example, the entire fore wing is leathery, in keeping with the sleek watertight design of their bodies. In shield or jewel bugs, the last section of the thorax (the scutellum) is expanded into a large leathery plate, which entirely covers both pairs of wings.

Typical bugs come in a wide variety of shapes, sizes and colours, ranging from glittery jewel bugs and brightly patterned assassin bugs to cryptically camouflaged water bugs. Regardless of their appearance, almost all members of this Suborder have well-developed compound eyes, a piercing mouth known as a rostrum and long thin antennae, which often kink outwards near the ends. In some aquatic forms, the antennae are tiny and slicked back against the head to make the body more streamlined for swimming.

In most typical bugs the legs are slender, equal in size and used for walking. However, some groups have interesting modifications to their legs to allow them to exploit different food types and habitats. Assassin bugs and giant water bugs have raptorial fore legs – strong limbs ending in a sharp claw to help them catch, kill and grasp their prey. Water striders have extremely long slender legs that enable them to balance on the water's surface, without toppling in. Backswimmers have flat oar-shaped hind legs covered in a fringe of tiny hairs to pull themselves through the water.

Diet and habitat

Most members of the Order Hemiptera live on land and eat a plant-based diet. However, members of the Suborder Heteroptera have broken both of these trends – they can be found in aquatic habitats and have broadened their palate to include the blood and tissue of other animals.

Admittedly, many shield bugs and plant bugs still prefer their greens and use their piercing–sucking mouthparts to remove the sap and juices from the stems, leaves, fruits and seeds of plants.

Others have developed a taste for blood. Bed bugs are pests of human dwellings, and these sneaky creatures siphon off our blood as we sleep, retreating into secret hidey-holes in bedrooms before we wake to catch them in the act. Thankfully, good hygiene practices such as vacuuming around the home keep bed bug infestations under control, but they still appear at times in some hotels and other public accommodations.

Assassin bugs, as the name suggests, are ruthless predators. These stealthy creatures creep up on unsuspecting insects and plunge their rostrum into their bodies. Saliva containing a potent cocktail of digestive and other enzymes is pumped into the prey, immobilising the insect and reducing its innards to a mushy soup. This is swiftly sucked back up through the rostrum, much like us drinking a milkshake through a straw.

'Water bugs' is a collective term given to the many typical bugs that have left dry land in favour of a life underwater. Some, such as water striders, merely get their toes wet, tip-toeing across the water's surface (held up by the surface tension) as they scavenge on the carcasses of insects that have fallen into the water and drowned. Others, such as giant water bugs, swim through the water with powerful strokes of their legs in search of fish and tadpoles to feast upon. Needle bugs and water scorpions (not really a scorpion, but a typical bug that resembles one) often go unnoticed, with brown bodies covered in algae that help them blend into their surroundings as they stalk other aquatic animals.

Lifecycle

Mate attraction in typical bugs often involves sound production, but it is much more subtle than the noisy songs of cicadas. Males produce sounds by stridulation, a process by which various parts of the body are rubbed together to produce a noise. Interestingly, different groups of bugs have their own preferred method of crooning to potential mates. Assassin bugs rub their mouthparts against the underside of their bodies, shield bugs swipe their legs over their abdomens, and jewel bugs knock their genitalia together to impress their lady friends. Once a suitable mate has been found, the pair often mates tail-to-tail, with the usually larger female dragging her puny partner around as she forages for food.

While most insects simply lay their eggs and leave, many typical bugs show a touching level of parental care. Jewel bugs and stink bugs lay clusters of up to 50 eggs, often glued to the surface of leaves or dotted around the stem of a plant. The females guard these eggs, manoeuvring their bodies over their clutch so that they are sheltered underneath their abdomens. Female giant water bugs place their eggs in the safest place available to them – on the backs of their dutiful spouses! The males swim around, eluding hungry predators and regularly resurfacing to expose the eggs to the air, which helps them to incubate.

As with all true bugs, typical bugs undergo incomplete metamorphosis. The young, called nymphs, resemble their parents, but are smaller, wingless and have shorter antennae. They shed their skins around five times during their journey to adulthood to accommodate their expanding bodies. Adults and nymphs often share habitats and diets and can be found feeding side by side.

Defence

The first line of defence for many typical bugs has earned them the nickname 'stink bugs'. Both adults and nymphs have a complex network of scent glands dotting their bodies, linked to a reservoir of pungent chemicals. At the slightest threat these glands release a repugnant smell, resulting in a hasty retreat by humans and predators alike. Some, such as the bronze orange bug, can squirt these noxious fluids for 20 cm or more, causing an irritating burning sensation upon contact.

Many typical bugs are well camouflaged to avoid detection by enemies. Some are convincing bark mimics, while others blend into leaves, flowers and seed pods. The nymphs of some pod-sucking bugs resemble small ants. Being mistaken for an ant has its benefits – many predators avoid ants as they often have sharp mandibles and painful stings. Others, such as assassin bugs and jewel bugs, are not well camouflaged and instead use their bright warning colours to advertise their distastefulness to predators such as birds and lizards.

Goodie or baddie?

Some typical bugs are good, some are bad. Although they do not obviously defoliate leaves like chewing beetles and grasshoppers, plant bugs such as mirids, green vegetable bugs and seed-suckers can inflict substantial damage on agricultural crops. They inject enzymes that break down plant tissue then they suck it up. This mode of feeding may stunt plant growth, reduce crop yield, encourage wilting and other deformities and potentially spread plant viruses. And, of course, bloodsucking bed bugs are a very unwelcome visitor in our homes.

By contrast, many typical bugs are predators and can help keep populations of pest insects under control. Assassin bugs are a great addition to your vegetable garden, efficiently picking off any caterpillars or grasshoppers that may be making a meal out of your plants and herbs. Water bugs are also highly beneficial insects as they consume vast quantities of mosquito wrigglers before they have a chance to grow into bloodsucking adults.

Fascinating facts

- Mirids, otherwise known as leaf bugs or plant bugs, have a reputation for being sneaky freeloaders. These crafty insects loiter around carnivorous plants such as sundews, which produce droplets of adhesive fluid on the surface of their leaves to capture prey. When a small insect becomes entangled the mirid moves in, tiptoeing around the sticky

This male giant water bug (*Diplonychus* sp.) carries a batch of eggs on his back, deposited there by the female. Source: Gooderham and Tsyrlin (2002).

INSECT ORDERS 297

The colourful nymphs of the banana-spotting bug (*Amblypelta lutescens*). Source: D. Papacek.

Feather-legged assassin bugs (*Ptilocnemus lemur*) are specialised predators of ants. When an ant crosses its path, the bug arches back, revealing a tiny droplet of delicious fluid on the underside of its body, which the ant greedily laps up. However, this secretion is laced with narcotics and the drugged ant collapses at the bug's feet and is quickly devoured. (Source: L. Woodmore).

snares to steal the struggling creature, depriving the sundew of its meal.
- Mating is a brutal ordeal for the female bed bug. She lacks a proper genital opening, so her mate simply uses his sharp reproductive organ to punch a hole into the side of her body to inseminate her.
- Water scorpions are aquatic typical bugs with pincer-like fore legs and a long tail, resembling a scorpion. But this tail is not for stinging – it is an elaborate butt-snorkel! Its design allows the insect to siphon oxygen from the surface while keeping its head underwater, so it can remain on the lookout for potential prey or stay vigilant against predators.

Learn more

In this book:

Assassin bugs, pp. 7, 109
Backswimmers, p. 98
Bed bugs, p. 84
Bronze orange bugs, pp. 22, 116
Brown bean bugs: adults, p. 123, ant-mimicking nymphs, p. 180
Cotton harlequin bugs, p. 108
Crusader bugs, p. 208
Giant water bugs, p. 95
Green vegetable bugs, p. 122
Gum tree shield bugs, p. 126
Shield bugs, p. 276
Spined citrus bugs, p. 117
Water scorpions and needle bugs, p. 97
Water striders, p. 96
Zebra gum tree shield bugs, p. 188

In other books:

See Further Reading

Online, type in:

Assassin bug ant corpse armour – a bizarre species of ant-eating assassin bug stacks the corpses of its recent meals on its back as a means of avoiding predators.
Beautiful scutelleridae – type this into an images search to see amazing examples of this Family of jewel bugs.

Non-insect arthropods

Insects belong to a much larger assemblage of animals known as arthropods, a group that includes spiders, scorpions, ticks, mites, millipedes, centipedes and crustaceans (see Chapter 2, pp. 5, 8). This section provides information on some of the non-insect arthropods commonly encountered in the home and garden.

The classification of arthropods is very complicated and constantly changing as scientists discover more about the biology and diversity of its members. Books on this subject are a minefield of confusing terminology such as Phyla, Classes, Subclasses and Orders, and just when you think you have it figured out you will find another book that uses a completely different ranking system. To make things easier, the following groups of non-insect arthropods are simply referred to by their common names – higher levels of classification have been avoided. Should you wish to learn more, simply follow the recommended reading list for non-insect arthropods in Further Reading.

Spiders

The main difference between spiders and insects comes down to their legs – spiders have eight legs, insects have six. Spiders have a body that is divided into two distinct segments, separated by a narrow waist. The head and the thorax are fused into a single part known as the cephalothorax (or *prosoma*) and the abdomen (or *opisthosoma*) is situated behind this segment.

Spiders have a small pair of appendages near their head known as palps that are often mistaken for antennae (which spiders lack). In male spiders, the tips of the palps are swollen like little boxing gloves; these are used during mating to transfer bundles of sperm to the body of the female. At the end of a spider's abdomen are a pair of

This giant golden orb weaver (*Nephila pilipes*) is as big as an adult's hand and builds extremely strong webs from strands of golden silk.

finger-like spinnerets, which are used to spin silk for ensnaring prey, wrapping up food or producing protective sacs for eggs.

Spiders are predators, hunting any animals that are small enough to overpower. Their diet usually consists of insects, but large spiders are known to eat lizards, frogs and even small birds! Spiders lack jaws and therefore cannot chew their food. Instead, they use powerful venom to kill their prey, then smother it with digestive enzymes to make it nice and mushy, before finally sucking the gooey mess into their mouth.

This wood scorpion (*Cercophonius* sp.) shelters under bark or leaf litter during the day. Source: J. Dorey.

Spiders can be found in a wide variety of habitats. Some are nocturnal, sheltering under rocks or bark or in underground burrows by day and emerging at night to wander the ground in search of food. Others build magnificent orb-shaped webs, ensnaring flying insects in their sticky strands. Some spiders are semi-aquatic, either striding on the water's surface or briefly diving underneath it to snare small fish and aquatic insects.

Australia is home to some of the world's most venomous spiders, including the red-back spider and the Sydney funnel-web. The introduction of anti-venoms to combat bites from these spiders has significantly reduced the number of spider-related fatalities.

Scorpions

Scorpions, like spiders, have eight legs but are quite different in appearance from their spider cousins. Their bodies are flattened and slender, with a pair of claws near their head for grasping prey and a long tail armed with a sharp sting at the tip.

Scorpions are nocturnal predators, emerging after dark to feast on a wide variety of animals including insects, other scorpions and small lizards. They grasp their food with their claws, while their tail arches up over their head and the stinger administers a fatal dose of venom to subdue the prey. They douse their food with digestive enzymes and suck out the liquid with their pincer-like mouthparts.

Most people associate scorpions with desert habitats and, indeed, many species occur in deep underground burrows in these arid areas. But scorpions can also be found in rainforests and woodlands, either sheltering under rocks or logs on the ground, or hidden under the bark of trees.

Scorpions perform an interesting dance as part of their mating ritual. The male deposits a tiny package of sperm on the ground. Grasping the claws of the female with his own, he skilfully manoeuvres her over it so that she can take it up into her body. Scorpions give birth to live young, with the tiny hatchlings hitching a ride on mum's back for a few weeks until they are big enough to fend for themselves.

Some species of scorpions around the world are known for their extremely powerful venom, with stings sometimes resulting in death. However, the stings of Australian scorpions are much milder and have been likened to that of a wasp. While scorpion stings can be very painful, fatalities from them in Australia are extremely rare.

Ticks and mites

Ticks and mites form a surprisingly diverse group of invertebrates. Like their close relatives the spiders and scorpions, they have eight legs and lack antennae. They are quite small in size, with even the largest of ticks barely reaching 10 mm long. Many mites are microscopic and cannot be seen with the naked eye.

Mites can be found almost everywhere, both on land and in fresh water. Some are predators, feeding on tiny insects, insect eggs and other species of mites. Others use their tiny piercing mouthparts to suck the fluid from the leaves and shoots of plants; these include a few species of mites that are known pests in agricultural crops. Ectoparasitic mites hitch a ride on the bodies of other animals (including humans), feeding on blood and other body fluids. Endoparasitic mites target the insides of their host, infesting the nasal cavities, lung chambers, ear canals and bowels of animals.

Ticks, on the other hand, are specifically adapted to feed on the blood of vertebrates. They target many types of animals including cattle, dogs and cats, kangaroos and wallabies, bandicoots, lizards, echidnas and some birds. They have a needle-like mouth to pierce flesh and rows of teeth to help anchor them to the body of their host. Their salivary glands release toxins into the victim, including special chemicals to speed the flow of blood and prevent it from clotting as they feed.

Ticks and mites lay eggs, from which hatch tiny larvae. The larvae have only six legs, but they gain an extra pair after they grow and moult their skin. Some species change to a different host with each moult, up to three times.

Some ticks and mites are of medical and veterinary importance. The scabies mite can cause severe rashes, allergic reactions and skin conditions in both humans and animals. The bloodsucking lifestyle of ticks can result in transmission of diseases such as tick typhus, and the toxins they inject may cause paralysis and even death in some animals.

This tick (*Ixodes* sp.) has long legs armed with claws to grasp onto the body of its host. Source J. Dorey.

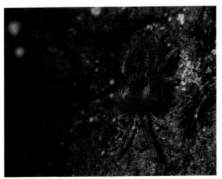

Red velvet mites (Family Trombidiidae) get their name from the dense coat of tiny setae that covers their bodies. Source: J. Dorey.

Centipedes

Despite their name (*centi* comes from a Latin word meaning hundred), centipedes rarely have 100 legs. Some centipedes have as few as 30 legs, others may have as many as 300. However, the centipedes we most commonly encounter here in Australia have fewer than 50 legs, usually between 42 and 46.

A large centipede (*Scolopendra laeta*). To the right is its head (with long antennae) and to the left are the pincers at the end of its abdomen. Source: L. Woodmore.

Centipedes have one pair of legs for each segment along their long, slightly flattened bodies. The first pair of legs near the head is modified into 'venom claws', sharp-tipped appendages that are capable of delivering a painful nip. The last pair of legs are larger and thicker than the rest and may be lined with sharp spines. These 'pincers' are used to help grasp prey and for defence.

Centipedes are voracious predators and most commonly hunt under the cover of darkness. Their long, highly sensitive antennae act as radars, sweeping the ground for suitable prey such as insects, frogs, small lizards and mice. Their venom claws inject fast-acting toxins, swiftly immobilising their prey. Their powerful jaws then rip their meal into smaller, more manageable-size pieces.

Male centipedes deposit a tiny bundle of sperm on the ground, sometimes spinning a soft pad of webbing to place it on. The female takes up this bundle into her body and uses it to fertilise her eggs. The eggs are laid in shallow burrows in soil and leaf litter. Some females cradle their eggs in their legs, protecting them from enemies until they hatch into miniature versions of their parents.

Many different types of centipedes are found in Australia and they can reach over 15 cm in length. Some are capable of inflicting a very painful bite if handled or disturbed, with the pain often persisting for several days.

Millipedes

Millipedes are quite different in appearance from centipedes. For starters, millipedes are strict vegetarians and so lack the long antennae, venom claws and powerful pincers that centipedes use to detect and capture their prey. Instead, millipedes have much smaller antennae that may be tucked into little grooves along the sides of their heads. Their cylindrical, sausage-like bodies lack large jaws and pincers and their movements are slower and more graceful than the fast aggressive strides of centipedes. Also, millipedes have two pairs of legs for each segment of their bodies, while centipedes have only one pair per segment.

The prefix 'milli' suggests these arthropods have 1000 legs, but the total number falls well short of this and can range anywhere from 26 to 700. Many types of millipedes are found in Australia. Short stout pill millipedes roll into tight little balls when disturbed, while rainforest millipedes have extremely long bodies that can reach up to 20 cm in length.

A juvenile millipede (cf. Order Spirobolida) showing its cylindrical body and numerous legs. Source: J. Dorey.

Millipedes prefer moist environments and can be found in leaf litter, soil and compost bins and under rocks and logs. They feed on a diet of decomposing plant material and often turn up in large numbers after heavy periods of rain.

Male millipedes bundle their sperm into little packages. A male locks with the female in a tight embrace and uses his legs to transfer the sperm into a special opening in the female's body. The eggs are laid in soil or under logs and rocks. The tiny hatchlings resemble their parents, but are much smaller and paler.

Millipedes, unlike centipedes, cannot bite. When provoked, some species release yellow or red fluid containing hydrogen cyanide from special openings along the sides of their bodies. This fluid is designed to make them distasteful to potential predators and may cause irritation if it comes into contact with your skin or eyes. It can also stain your skin red or brown (often taking several days to fade completely) and can stain clothes permanently.

Slaters

Also known as woodlice and pillbugs, slaters are a group of tiny crustaceans which, unlike their shrimp and crayfish cousins, live on dry land. Slaters have 14 legs and two pairs of antennae, although the second pair can be quite hard to see. Slaters grow up to 15 mm long and have heavily armoured, oval-shaped bodies that are usually grey. If disturbed, some slaters can roll their bodies into a tight defensive ball, much like an armadillo.

Slaters can be found in moist habitats such as soil and leaf litter and under rocks and logs on the forest floor. They are

Garden slaters (Order Isopoda) are a type of land-dwelling crustacean, commonly found in soil, leaf litter and compost. Source: J. Dorey.

frequent visitors to our vegetable gardens and compost bins. Most species feed on decaying vegetable material and fungi. However, some species are scavengers and include dead animal matter in their broad diets.

Male slaters pass tiny packages of sperm to the female using their last pair of legs. The female uses this sperm to fertilise up to 100 eggs, which she keeps tucked in a special pouch underneath her body until they hatch. The tiny hatchlings resemble their parents, but have just 12 legs instead of 14. They gain the extra pair as they grow and moult their skins.

Some gardeners have reported slaters feeding on living plants such as young seedlings, shoots and fallen fruit, in their vegetable gardens. However, the damage they cause is considered to be quite minor. Slaters are totally harmless to humans and play an important role in breaking down and recycling nutrients in our gardens.

Land hoppers

Land hoppers, like slaters, belong to a group of crustaceans that live on land rather than water. Their tiny bodies are tall and skinny, much like a flea. They are usually less than 10 mm long and are tan, brown or grey.

Land hoppers (Order Amphipoda), as their name suggests, leap into the air when disturbed. Source: J. Dorey.

Land hoppers have 26 limbs in total, running along the segments of their body. They have seven pairs of walking legs, three pairs of shortened legs and three pairs of tail legs, directed backwards towards the end of their bodies. Despite all these limbs, land hoppers are quite clumsy when it comes to walking. Their real talent lies in their ability to jump.

When disturbed, land hoppers suddenly flick their abdomen backwards, launching their bodies into the air. They are found in moist environments such as damp leaf litter and soil and under flower pots and logs, where they feed on decaying plant material. We usually notice land hoppers only when we disturb them and see them leaping about as we shift pots, rake up leaves and move rocks and logs.

After mating with a male, the female land hopper houses her eggs in a special brood pouch on the underside of her body until they hatch. The young moult their skins several times to reach adulthood.

Land hoppers are harmless to humans and play an important role in recycling nutrients in our gardens.

Glossary

This guide provides definitions for some of the terms used in this book.

sing. = the singular form of a word.

pl. = the plural form of a word.

Abdomen – the last of the three segments of an insect's body. The abdomen is at the rear end of the insect and contains the digestive and reproductive systems and much of the nervous system.

Adult – the final stage in the life of an insect by which point growth, development of the reproductive organs and the formation of wings is complete. With the exception of silverfish, all insects stop moulting their skin once they have become adults and do not grow any larger.

Alates – a special group of winged fertile males and females found in colonies of social insects such as ants and termites. Their job is to disperse from the nest, mate and start new colonies.

Ametabolous metamorphosis – *see* **Simple metamorphosis**.

Antennae (sing. antenna) – a pair of sensory structures found on either side of an insect's head, used to sense touch, taste and smell.

Anting – a behaviour seen in birds, in which the bird picks up an ant in its beak and rubs it over its plumage, or allows ants to crawl over its spread wings. The secretions produced by the ants presumably discourage infestations of lice.

Arachnid – an organism belonging to a Class of arthropods known as Arachnida, which includes spiders, scorpions, ticks and mites. Arachnids have eight legs and two body parts (cephalothorax and abdomen), and lack antennae.

Arolium – a cushion-like pad located between the claws on the tarsi (feet) of some insects. It allows the insect to grip onto slippery surfaces.

Arthropod – an organism belonging to the Phylum Arthropoda, which includes spiders, scorpions, millipedes, centipedes, insects and crustaceans. Arthropods have a hard exoskeleton and jointed legs.

Beneficial insect – an insect that provides a service that positively impacts on the lives of humans. Insect predators or parasites that help to control a pest plant or animal species, and insects that pollinate plants, are examples of beneficial insects.

Biological control – the deliberate use of a living organism (e.g. an insect) to control populations of pests (animals or plants), for example releasing predators such as green lacewing larvae into orchards to control plant-sucking mealybugs.

Bristle-like antennae – antennae that are reduced to a pair of small hair-like bristles, often difficult to see with the naked eye. Flies, dragonflies and cicadas are example of insects with bristle-like antennae.

Bristles – short stiff hairs on the body of an insect.

Bug – a term used by many people as a substitute for 'insect'. However, a true bug is an insect belonging to the Order Hemiptera and has piercing–sucking mouthparts.

Camouflage – a technique used by insects to avoid detection by enemies by blending into their surroundings. Insects often look like leaves, sticks and rocks to aid in their camouflage.

Carnivore – *see* **Predator**.

Caste – a group of individuals of similar appearance in a colony of social insects and with a particular role within the nest, for example in termite colonies the soldier caste differs from the worker caste by having individuals with larger heads, strong mandibles and bigger, stronger bodies for defending the nest.

Caterpillar – a long cylindrical larva with a well-developed head and mandibles. This name is used for the larvae of butterflies and moths.

Cellulose – a type of carbohydrate used to construct the cell walls of plants. This tough material is the main component in wood and vegetation and no insects (except silverfish) can digest it by themselves. Some insects, though, have special organisms living in their gut that secrete enzymes that can break down cellulose.

Cenchri – rough bumps on the thorax of sawflies. These are used to produce sounds to startle predators.

Cephalothorax – the first of the two segments of an arachnid's body, made up of the fused head and thorax.

Cerci (sing. cercus) – a pair of finger-like sensory structures found at the end of the abdomen in some insects. Cerci are used to pick up signals from the surrounding environment, to assist in mating, or for defence.

Cf. – an abbreviation derived from the Latin word *confer*. In taxonomy, cf. is placed between the Genus and Species names if the identity of the specimen is uncertain, either because the specimen was poorly preserved making identification difficult or, in the case of photographed specimens, if not all key diagnostic features can be seen clearly. In this book, a photograph of a cockroach labelled *Periplaneta* cf. *americana* implies that the cockroach in question belongs to the Genus *Periplaneta* and is likely the Species *americana*, but because the angle of the photograph obscures important diagnostic features there is a level of uncertainty.

Chrysalis (pl. chrysalises) – the special name given to the pupa of a butterfly.

Class – a classification ranking of organisms that sits between Phylum and Order. A Class is made up of one or more Orders, for example Class Insecta includes around 30 different Orders of insects.

Clubbed antennae – antennae that are gradually thickened at the tip farthest from the head. For example, butterflies have clubbed antennae, moths do not.

Cocoon – a protective case or bag produced by some types of insect larvae to surround and protect their pupa. It is usually made of silk produced by the larva and may have leaves, itchy hairs or frass woven into it to aid in concealment.

Colony – a group of individuals belonging to a single species of insect that live and cooperate in a communal nest or hive.

Complete metamorphosis – a type of development seen in modern winged insects such as butterflies, beetles, bees, wasps and ants. In this type of lifecycle, the insect passes through four distinct stages – the egg, larva, pupa and adult. The hatchling insect looks like a grub or caterpillar and does not resemble the adult at all. Wings and reproductive organs are developed within a pupa.

Compound eyes – a pair of visual organs found on an insect's head. Each compound eye is made up of many (up to several thousand) small lenses called facets that join together to form the eye.

Coxa (pl. coxae) – the uppermost part of an insect's leg that attaches to the thorax. It is similar to the hip joint of a human.

Crepuscular – used to describe organisms that are most active at dusk or dawn.

Crochet – the tiny hooks on the prolegs of caterpillars.

Cryptic – used to describe organisms that are hidden, secretive or difficult to detect.

Cuticle – a slippery or waxy layer covering the outside of an insect's body. It helps to keep the insect's exoskeleton waterproof.

Defoliate – to completely strip the leaves from a plant.

Detritus – disintegrated decomposing material such as mulch, leaf litter and decomposing vegetation.

Dichotomous key – a tool used by scientists to identify things. It consists of pairs of questions about the physical appearance of a given organism. Each answer discards possibilities, until a single answer (the identity of the organism) remains.

Diurnal – used to describe organisms that are most active during the day.

Dorso-ventrally flattened – describes an organism with a body flattened between its upper and lower surface. In other words, it describes things that are flat, like lice.

Ecdysis – *see* **Moulting**.

Ectoparasite – an organism that lives and feeds on the outside of another animal's body, without killing it. Fleas, lice and ticks are examples of ectoparasites and can be found on the feathers, fur or skin of their hosts.

Egg – a round or oval-shaped capsule surrounding the developing embryo of an organism. The egg is the first stage of an insect's development.

Elbowed antennae – antennae that are bent at a 90° angle in the middle. Ants have elbowed antennae.

Elongate – long and slender, often used when describing the bodies of insects such as praying mantids and stick insects.

Elytra (sing. elytron) – the hardened fore wings of beetles and earwigs. They form a close-fitting cover that sits tight and flat over the delicate hind wings and abdomen.

Empodium – a bristle-like structure located between the claws on the tarsi (feet) of insects such as flies. It is used to help grip onto slippery surfaces.

Endoparasite – an organism that lives and feeds inside the body of another animal, without killing it. Some mites are endoparasites, living in the ear canals, nasal cavities and lung chambers of their host.

Entomologist – a scientist who studies insects.

Entomology – the study of insects.

Eversible – a structure that can be turned inside out by an organism.

Exoskeleton – the tough external skeleton that surrounds an arthropod's body. It protects against dehydration and extreme environmental conditions.

Eye spots – circular markings on the bodies (often the wings) of insects that resemble a giant pair of eyes. This can help insects trick their enemies into thinking they are a much larger animal.

Facet – the individual lens of a compound eye.

Family – a classification ranking of insects that sits between Order and Genus. A Family is made up of one or more Genera and may be further divided into Subfamilies.

Feathery antennae – antennae that are covered on either side by slender projections, making them look like a feather. Moths have feathered antennae.

Femora (sing. femur) – the section of an insect's leg between the coxa and the tibia. It is similar to the thigh of a human.

Forceps – a strong pair of curved appendages at the end of an earwig's abdomen. They are more heavily armoured and rigid than cerci.

Fore leg – the first of the three pairs of an insect's legs, closest to the head.

Fore wing – the first of the two pairs of an insect's wings, closest to the head.

Frass – the technical name for insect poo, usually used to describe the pellet-like droppings of insects such as caterpillars.

Gall – an abnormal tumour-like swelling of plant tissue often caused by the feeding action of certain insects.

Gaster – the name given to the main segment of an ant's abdomen, located behind the narrow waist.

Genus (pl. Genera) – a classification ranking of insects that sits between Family and Species. A Genus is made up of one or more Species. The Genus name is always written in italics and begins with a capital letter.

Gills – respiratory organs evident in some aquatic insects, allowing them to breathe underwater.

Grub – the name usually given to the larvae of beetles.

Habitat – in this book, the word habitat refers to the location where an insect naturally lives (e.g. mosquito wrigglers in ponds), feeds (e.g. swallowtail caterpillars on citrus plants) or otherwise typically occurs (e.g. moths attracted to artificial lights).

Halteres – the highly modified hind wings of flies, which are reduced to a tiny pair of clubs or hairs.

Hamuli – a row of tiny hooks found on the wings of wasps (and some other insects, e.g. aphids) that allow them to join their fore and hind wings together during flight.

Head – the first of the three segments of an insect's body. The head is at the top of the body and bears the eyes, antennae and mouthparts.

Head capsule – a series of hardened plates on the head of an insect, fused into a solid capsule.

Hemelytra – the leathery fore wings of true bugs belonging to the Suborder Heteroptera. They form a close-fitting cover that sits tight and flat over the delicate hind wings and abdomen and consist of a thickened base and a membranous tip.

Hemimetabolous metamorphosis – see **Incomplete metamorphosis**.

Herbivore – an animal that eats only plant material.

Hind leg – the last of the three pairs of an insect's legs, closest to the abdomen.

Hind wing – the second of the two pairs of an insect's wings, closest to the abdomen.

Holometabolous metamorphosis – see **Complete metamorphosis**.

Honeydew – a waste product in the form of a sweet liquid, which is expelled from the bodies of insects (e.g. aphids and cicadas) that feed on plant fluids.

Host – the organism upon which a parasite lives and feeds.

Incomplete metamorphosis – a type of development typical of primitive winged insects such as stick insects, praying mantids, cockroaches, cicadas and dragonflies. In this type of lifecycle, the newly hatched insect resembles the adult, but is smaller and lacks wings. The insect simply grows larger with each successive moult and the wings develop in external sheaths on the insect's back.

Insect – an organism belonging to a Class of arthropods known as Insecta, which includes (among others) butterflies, beetles, stick insects and grasshoppers. Insects have six legs, three body parts (head, thorax and abdomen) and a single pair of antennae.

Instar – the name given to the active immature stages of an insect's life, between the egg stage and the pupal or adult stages. For example, a caterpillar that has hatched out of its egg is known as a first instar; it enters the second instar after it moults its skin for the first time. The final instar in a caterpillar is the one before it changes into a pupa.

Invertebrates – the term used to describe all animals that lack a true backbone. As well as insects and their close relatives the spiders and scorpions, invertebrates include jellyfish, worms, snails, starfish and barnacles.

Johnston's organ – a sensory organ used to detect sound, located in the second segment of the antennae in insects such as flies.

Jointed legs – the flexible legs of arthropods divided into distinct segments, held together by a series of bendable joints.

Kingdom – the most general classification ranking of an organism. All animals belong in the Kingdom Animalia.

Larvae (sing. larva) – the second (between the egg and the pupa) of the four stages in the lifecycle of an insect that undergoes complete metamorphosis. Also referred to as grubs, maggots or caterpillars, depending on the insect involved. Also see **Nymph**.

Laterally compressed – describes an organism that is compressed from either side of its body. In other words, it describes things that are relatively tall and skinny, such as fleas.

Legs – the flexible limbs of an insect, used for running, jumping, walking, swimming, climbing or digging. Three pairs of legs (fore, mid and hind) attach to the thorax.

Leks – an aggregation or swarm of male insects formed in an area where they attract and compete for the attention of females of their kind.

Lichen – a group of moss-like plants, usually greenish-grey, commonly found growing on the surface of rocks and trees.

Maggot – the name usually given to the larvae of flies.

Mandibles – the chewing appendages in the jaws of insects, used for biting and grinding food.

Mandibulate soldier – a special form of soldier termite with an enormous armoured head and massive jaw, used in defence.

Membranous – thin and transparent, often used when describing the wings of insects such as dragonflies and wasps.

Metamorphosis – the process by which a juvenile insect changes into an adult. In other words, metamorphosis sums up the developmental changes that happen to an insect throughout its life.

Mid leg – the middle of the three pairs of an insect's legs, located between the fore and hind leg.

Mimicry/to mimic – the art of pretending to be something you are not, often to avoid being spotted by a predator. For example, insects mimic bird droppings (not very appetising to a predator) or animals that are dangerous to predators (e.g. harmless flies mimic stinging wasps).

Minibeast – a non-scientific term that describes any small critters in our home or garden. While some people use the term minibeast as a substitute for arthropod, others lump invertebrates such as snails, slugs, leeches and worms under the title.

Modern insects – a term used to describe Orders of insects that have appeared more recently in relative terms (which may be many millions of years). These groups include insects that undergo complete metamorphosis, such as beetles and butterflies.

Morphology – the study of the structure and form of an organism.

Moulting – the complex process by which insects peel off their exoskeleton to reveal a new larger one underneath (much like a snake shedding its skin). This enables the insect to grow.

Mouthparts – insect mouthparts are complex, made up of different (usually paired) appendages.

Mutualism – a relationship between two organisms that is mutually beneficial to both.

Myrmecophile – literally meaning 'ant-loving'. A myrmecophile is an organism that lives in close association with ants and their nests, usually gaining food or shelter from them.

Myxomatosis – a viral disease of rabbits introduced to Australia to control populations of European rabbits. The virus is spread by bloodsucking organisms such as fleas and lice.

Naiad – the name given to the aquatic juveniles of insects such as dragonflies and damselflies. Despite undergoing incomplete metamorphosis, these naiads have adaptations (e.g. gills) to help them live underwater, making them look very different from their terrestrial parents.

Nasute soldier – a special form of soldier termite that uses a tubular opening on its forehead to shower enemies with toxic secretions. These soldiers have reduced mouthparts, so are different from mandibulate soldiers in the same nest.

Nocturnal – used to describe organisms that are most active at night.

Nuptial flight – where male and female insects take to the air to mate. Typical of social insects such as ants and termites.

Nymph (pl. nymphs) – the name given to the immature stage (between the egg and the adult) of an insect that undergoes incomplete metamorphosis. Sometimes also called a larva, but the latter is typically reserved for insects that undergo complete metamorphosis.

Ocelli (sing. ocellus) – small circular or oval-shaped sensory organs on the foreheads of many insects. These 'simple eyes' can help an insect to detect levels of light and darkness.

Omnivore – an animal that eats any kind of organic material, whether it is plant or animal in origin.

Ootheca (pl. oothecae) – a hardened or spongy protective case that surrounds a batch of insect eggs. Produced by insects such as praying mantids and cockroaches.

Opaque – not see-through, often used when describing the fore wings of insects such as grasshoppers and stick insects.

Opisthosoma – also referred to as the abdomen, this is the last of the two segments of a spider's body.

Order – a classification ranking of insects that sits between Class and Family. An Order is made up of one or more Families and may be further divided into Suborders.

Osmeteria – a pair of eversible glands on the thorax of swallowtail butterfly caterpillars. These horn-like structures feature bright colours and strong smells and are used to startle and scare away predators such as birds.

Ovipositor – a special egg-laying tube located at the end of the abdomen in many female insects. It can be serrated, needle-like or blade-like.

Palps – a pair of segmented finger-like projections attached to the mouthparts that sit near the mandibles of insects.

Parasite – an organism that lives and feeds on or in the body of another animal. Unlike parasitoids, parasites generally do not kill their host, but sometimes harm it indirectly by spreading pathogens which may affect its behaviour, health and reproduction.

Parasitise – to infect a host with a parasite. For example, many wasps parasitise caterpillars.

Parasitoid – an insect whose larvae live as parasites on or in the bodies of other organisms, eventually killing their host.

Parthenogenesis – a form of reproduction in which viable eggs are produced without fertilisation by a male.

Pest insect – an insect that negatively impacts on the quality and yield of agricultural crops, the integrity of our foodstuffs, clothing, furnishings and houses, or the health and safety of humans and domestic animals.

Petiole – the name given to the narrow first (and sometimes second) segment of the abdomen in insects with a constricted 'waist' (i.e. ants and wasps).

Pheromone – a chemical smell produced by insects used for communication with other members of their species (commonly in mating).

Phloem – a type of tissue in plants responsible for transporting sugars and nutrients from the leaves down through the plant stem.

Phylum – a classification ranking of insects that sits between Kingdom and Class. A Phylum is made up of one or more Classes.

Pincers – see **Forceps**.

Pollen basket – a hollow area surrounded by stiff hairs on the hind legs of some bees, used to transport pollen.

Pollination – the process by which pollen is transferred from the male to the female reproductive structures of a flower, for the purpose of fertilisation.

Polyembryony – a type of reproduction where more than one embryo develops from a single fertilised egg.

Predator – an animal that kills and eats other animals.

Pre-pupa – a special intermediate 'resting stage' in insects such as thrips, where significant structural changes occur in their development. The insect that enters into the pre-pupa resembles the adult in appearance, so the process is quite different from that of a larva entering a pupa, such as in the lifecycle of a butterfly.

Primitive insects – a term used to describe Orders of insects that were on our planet long before the appearance of Modern insects. It refers to insects that undergo incomplete metamorphosis (e.g. grasshoppers and cockroaches) or simple metamorphosis (e.g. silverfish).

Proboscis – a long, straw-like tube used for sucking up liquids such as nectar, as in butterflies and moths.

Prolegs – suction-cup-like pads found on the abdomen of some caterpillars to help them grip onto plants. Many people mistake these for legs, but unlike true legs (which are located on the thorax) prolegs are not jointed and they attach to the abdomen.

Prosoma – another name for the cephalothorax of spiders.

Protozoans – a group of single-celled organisms, some of which live in the digestive system of insects such as cockroaches and termites, where they secrete enzymes that help to break down the tough cellulose in plant material.

Pupae (sing. pupa) – the third (between the larva and the adult) of the four stages in the lifecycle of an insect that undergoes complete metamorphosis. During this inactive phase, the insect transforms from a larva into an adult.

Puparium – a protective, barrel-shaped capsule in which some flies pupate. It is made from the last larval skin of the maggot.

Pupate – the process by which a larva turns into a pupa.

Raptorial – describes legs that are lined with sharp spines to seize and grasp prey. Praying mantids have raptorial fore legs.

Rostrum – the name given to the needle-like mouthparts of insects belonging to the Order Hemiptera. It consists of sharp hollow tubes, bundled together in a sheath.

Scales – microscopic, flattened overlapping plates on the wings, body and limbs of butterflies and moths that give them a furry or velvety appearance.

Scavenger – an organism that feeds on dead or decaying plant and animal matter, such as rubbish, rotting fruit and animal carcasses.

Scopa – specialised hairs on the body of bees, used to collect and transport pollen.

Scutellum – a hardened shield-like plate located on the backs of insects such as true bugs.

Segment – the name given to the sub-divisions of an insect's body or limbs.

Setae (sing. seta) – special hairs on the body of an insect that can be used to smell, taste, touch or hear.

Simple metamorphosis – a type of development seen in primitive wingless insects such as silverfish. In this type of lifecycle, the insect emerges from its egg looking like a smaller version of the adult and moults its skin several times to reach sexual maturity. It continues to moult throughout its adult life.

Siphon – a tube-like structure at the tip of the abdomen in some aquatic insects, used for breathing.

Social insects – insects that live and work together with other members of their species in a communal nest or colony. Termites, ants, bees and wasps are examples of social insects.

Solitary insects – insects that are not social (i.e. do not live or work with other members of their species).

Species – a taxonomic category composed of organisms that are capable of interbreeding. In the classification ranking of insects, Species sits below Genus. The Species name is always written in italics and begins with a lowercase letter.

Spermatheca – a reservoir in the bodies of female insects in which sperm from the male is stored.

Spermatophore – a capsule that contains bundles of sperm produced by insects such as silverfish.

Spinnerets – a small pair of finger-like appendages at the end of a spider's abdomen, used for spinning silk.

Spiracles – special holes (usually on the side of an insect's abdomen and thorax) used for breathing.

Sting/stinger – a tiny hardened needle at the end of the abdomen, used to inject venom into other animals. Many bees, wasps and ants have stingers.

Stridulation/to stridulate – the process by which insects make sounds by rubbing or clicking parts of their bodies together. For example, rhinoceros beetles rub their wing covers over their abdomens to make a hissing sound to startle enemies.

Stylet – a piercing tube in the mouthparts of insects such as true bugs.

Symbiotic organisms – organisms that form a close relationship with members of another species.

Tarsus (pl. tarsi) – the bottommost section of an insect's leg, located below the tibia. It is similar to the foot of a human.

Taxa – groups or units of classification formally recognised by taxonomists, consisting of organisms that show shared structural characters. Species, Genus, Family and so on are all taxa.

Taxonomist – a scientist who studies taxonomy.

Taxonomy – the scientific practice of classifying, identifying and naming organisms.

Tegmina (sing. tegmen) – the thick leathery fore wings of insects such as cockroaches, crickets, grasshoppers and praying mantids.

Terrestrial – describes something that lives on land.

Thorax – the middle of the three segments of an insect's body, found between the head and the abdomen. The legs and wings join to the thorax, which contains the muscles needed to operate them.

Thread-like antennae – antennae that are extremely long (often longer than the body of the insect) and thin, like a thread or a whip. Cockroaches and crickets are examples of insects with thread-like antennae.

Tibia (pl. tibiae) – the section of an insect's leg between the femur and the tarsus. It is similar to the shin of a human.

Trachea – a tube in the body of an insect responsible for the transportation of oxygen around the body.

Trochanter – the section of an insect's leg between the coxa and the femur. It works as a flexible joint to aid in leg motion.

True bug – an insect belonging to the Order Hemiptera, which has piercing–sucking mouthparts.

Tymbals – a thin membrane on the abdomen of cicadas, used to produce sound.

Tympana (sing. tympanum) – hearing organs found on the abdomen or fore legs of grasshoppers, cicadas and some moths. It is composed of a thin membrane that is sensitive to vibrations.

Vein – a hollow tube-like structure that provides structural support in the wings of insects.

Venom – toxic substances produced by special glands within the body of insects and some arachnids. Venom is injected into other organisms through the mouth or stinger and is used to kill prey or as a means of defence.

Wing – an appendage attached to the thorax of many insects, usually used for flight.

Wing buds – an external sheath on the thorax of insects that undergo incomplete metamorphosis, from which the adult wings develop.

Xylem – the woody tissue in plants responsible for transporting water and nutrients from the roots to the leaves.

Pronunciation guide

Many words used in the study of insects have Greek or Latin origins, making it difficult to know how to pronounce them correctly. This guide will help you wrap your tongue around some of the terms and names used throughout this book.

Abdomen – ab-doh-men

Alates – al-aytes

Ametabolous metamorphosis – ay-metab-o-luss meh-ta-more-foe-sis

Anisoptera – an-eye-sop-terra

Antenna – an-ten-na

Antennae – an-ten-nay (it is also commonly pronounced an-ten-nee)

Apocrita – ap-o-cry-ta

Arachnid – ar-rack-nid

Archaeognatha – ark-ee-og-nay-tha

Arolium – a-roll-ee-um

Arthropods – arth-roe-pods

Auchenorrhyncha – or-ken-noe-rin-ka

Auchenorrhynchans – or-ken-noe-rin-kans

Blattodea – bla-toe-dee-a

Brachyptera – bra-kip-terra

Caelifera – see-liff-erra

Caste – carst

Cenchri – sen-chree

Cephalothorax – seff-fallow-thorax (it is also commonly pronounced keff-fallow-thorax)

Cerci – serr-see

Cercus – serr-sus

Chrysalis – chris-a-liss

Cicada – sick-ar-da (also commonly pronounced sick-ay-da)

Cocoon – cook-coon

Coleoptera – koll-lee-op-terra

Collembola – koll-em-bolla

Coxa – cocks-za

Coxae – cocks-zee

Crepuscular – crep-uss-cue-lar

Crochet – crotch-et

Cryptic – krip-tick

Cuticle – cue-tee-cal

Dermaptera – der-map-terra

Detritus – der-try-tuss

Dichotomous key – die-kot-toe muss key

Diplura – die-ploo-ra

Diptera – dip-terra

Diurnal – die-urn-al

Ecdysis – ek-die-sis

Ectoparasite – ek-toe-para-site

Elytra – eee-lye-tra

Elytron – eee-lye-tron

Embioptera – em-bee-op-terra

Empodium – em-poe-dee-um

Ensifera – en-siff-erra

Entomologist – en-toe-moll-oh-jist

Entomology – en-toe-moll-oh-jee

Ephemeroptera – eee-fem-mor-op-terra
Epiproctophora – eppee-prock-toe-forra
Exoskeleton – ek-so-skell-ett-on
Femora – fee-more-ah
Femur – fee-murr
Forceps – fore-seps
Formicidae – form-miss-id-ee
Gall – gaul
Gaster – gass-ter
Genus – jee-nuss
Halteres – holt-airs
Hamuli – ham-you-lye
Hemelytra – heem-meh-lye-tra
Hemimetabolous metamorphosis – hem-mee-metab-o-luss meh-ta-more-foe-sis
Hemiptera – hem-mip-terra
Heteroptera – het-er-op-terra
Heteropterans – het-er-op-terans
Holometabolous metamorphosis – hollow-metab-o-luss meh-ta-more-foe-sis
Hymenoptera – hy-men-op-terra
Invertebrate – in-ver-teb-rate
Isoptera – eye-sop-terra
Katydids – kay-tee-dids
Larva – lar-va
Larvae – lar-vay
Leks – lex
Lepidoptera – leppy-dop-terra
Mandibles – man-dib-ells
Mantodea – man-toe-dee-a
Mecoptera – mee-kop-terra

Megaloptera – mega-lop-terra
Membranous – mem-bran-uss
Metamorphosis – meh-ta-more-foe-sis
Midges – mid-jez
Morphology – more-follow-jee
Moulting – molt-ing
Myrmecophile – mer-mee-co-file
Myxomatosis – mix-oh-mat-toe-sis
Naiads – ny-adds
Nasute – nor-suit
Nematocera – nemma-toss-serra
Neuroptera – new-rop-terra
Nocturnal – nock-turn-al
Ocelli – oh-sell-ee
Odonata – o-do-narta
Ootheca – oh-oh-thee-ka
Oothecae – oh-oh-thee-kee
Opisthosoma – oh-piss-tho-so-ma
Orthoptera – or-thop-terra
Osmeteria – oz-mah-teer-ree-ah
Ovipositor – oh-vee-pozz-ee-tor
Parasitic – para-sit-ick
Parthenogenesis – parth-enno-jenn-nis-sis
Petiole – pee-tee-ol
Phasmatodea – faz-ma-toe-dee-a
Pheromone – fair-rem-moan
Phloem – flow-em
Phthiraptera – theer-ap-terra
Phylum – fy-lum
Pincers – pin-sers

Plecoptera – plee-kop-terra
Polyembryony – poly-em-bree-onee
Proboscis – pro-boss-kiss
Prolegs – pro-legs
Prosoma – pro-so-ma
Protozoans – pro-toe-zoh-ans
Protura – pro-tue-ra
Pseudoscorpion – siyoo-doe-scor-pee-on
Psocids – soe-kids
Psocoptera – soe-kop-terra
Psyllids – sill-ids
Pupa – peeyou-pa
Pupae – peeyou-pay
Puparium – peeyou-pare-rhee-um
Pupate – peeyou-pate
Raptorial –rhapt-torr-ee-all
Rostrum – ross-strum
Scopa – scoe-pa
Scutellum – skew-tell-um
Seta – see-ta
Setae – see-tee
Siphonaptera – sy-fonn-app-terra
Species – spee-sees
Spermatheca – sperm-ma-thee-ka
Spermatophore – sperm-matto-four

Spiracles – spear-rah-cals
Sternorrhyncha – stern-noe-rin-ka
Sternorrhynchans – stern-noe-rin-kans
Strepsiptera – strep-sip-terra
Stridulate – strid-you-late
Stridulation – strid-due-lay-shun
Stylet – sty-let
Symbiotic organisms – symm-buy-otick organ-is-ims
Symphyta – sim-fy-ta
Tarsi – tarr-see
Tarsus – tarr-suss
Tegmina – tegg-mine-ah
Thorax – thor-ax
Thysanoptera – thy-san-op-terra
Thysanura – thy-san-new-ra
Tibia – tibb-bee-ya
Tibiae – tibb-bee-yee
Trichoptera – try-kop-terra
Trochanter – tro-kan-terr
Tymbals – tim-balls
Tympana – timm-pan-na
Tympanum – timm-pan-num
Xylem – zy-lum
Zygoptera – zy-gop-terra

Bibliography

The following books and journal articles were used in the research for this book.

Beccaloni J (2009) *Arachnids*. CSIRO Publishing, Melbourne.

Bell WJ, Adiyodi KG (1982) *The American Cockroach*. Chapman and Hall, London.

Blanche R (2012) *Life in a Gall: The Biology and Ecology of Insects that Live in Plant Galls*. CSIRO Publishing, Melbourne.

Braby MF (2004) *The Complete Field Guide to Butterflies of Australia*. CSIRO Publishing, Melbourne.

Broadley R, Thomas M (Eds) (1995) *The Good Bug Book: Beneficial Insects and Mites Commercially Available in Australia for Biological Pest Control*. Australasian Biological Control, Queensland Department of Primary Industries/Rural Industries Research and Development Corporation.

Brock PD (1999) *The Amazing World of Stick and Leaf-Insects*. Amateur Entomologists' Society, Kent.

Brock PD, Hasenpusch JW (2009) *The Complete Guide to Stick and Leaf Insects of Australia*. CSIRO Publishing, Melbourne.

Brooks S (2002) *Dragonflies*. Natural History Museum, London.

Brunet B (2000) *Australian Insects: A Natural History*. Reed New Holland, Sydney.

Burrows M (2006) Jumping performance of froghopper insects. *Journal of Experimental Biology* **209** (23), 4607–4621.

Burwell C (2007) *Ants of Brisbane*. Queensland Museum, Brisbane.

Burwell C (2008) *Backyard Insects of Brisbane*. Queensland Museum, Brisbane.

Burwell C, Monteith GB, Wright O, Peters BC (2007) Insects. In *Wildlife of Greater Brisbane*. 2nd edn, pp. 85–173. Queensland Museum, Brisbane.

Cawdell-Smith AJ, Todhunter KH, Perkins NR, Bryden WL (2012) Equine amnionitis and fetal loss: placentitis following exposure of mares to processionary caterpillars. *Equine Veterinary Journal* **44** (3), 282–288.

Chapman AD (2009) *Numbers of Living Species in Australia and the World*. 2nd edn. Department of the Environment, Water, Heritage and the Arts, Canberra.

Clyne D (2009) *The Secret Life of Caterpillars*. New Holland, Sydney.

Clyne D (2010) *All About Ants*. New Holland, Sydney.

Clyne D (2011) *Attracting Butterflies to Your Garden*. New Holland, Sydney.

Common IFB (1990) *Moths of Australia*. Melbourne University Press, Melbourne.

Common IFB, Waterhouse DF (1981) *Butterflies of Australia*. CSIRO Publishing, Melbourne.

Copeland M (2003) *Cockroach*. Reaktion Books, London.

Cornwell PB (1968) *The Cockroach: Volume 1.* Hutchinson & Co., London.

Coupar P, Coupar M (1992) *Flying Colours: Common Caterpillars, Butterflies and Moths of South-eastern Australia.* UNSW Press, Sydney.

CSIRO Division of Entomology (Eds) (1996) *The Insects of Australia: A Textbook for Students and Research Workers.* Melbourne University Press, Melbourne.

Dingle H, Zalucki MP, Rochester WA (1999) Season-specific directional movement in migratory Australian butterflies. *Australian Journal of Entomology* **38**, 323–329.

Dollin A, Batley M, Robinson M, Faulkner B (2000) *Native Bees of the Sydney Region: A Field Guide.* Australian Native Bee Research Centre, North Richmond.

Edwards R, Mill AE (1986) *Termites in Buildings: Their Biology and Control.* Rentokill, West Sussex.

Evans AV, Bellamy CL (2000) *An Inordinate Fondness for Beetles.* University of California Press, London.

Evans DL, Schmidt JO (1990) *Insect Defenses: Adaptive Mechanisms and Strategies of Prey and Predators.* State University of New York Press, Albany.

Field RP (2013) *Butterflies: Identification and Life History – Museum Victoria Field Guide.* Museum Victoria Publishing, Melbourne.

Floater GJ, Zalucki MP (1999) Life tables of the processionary caterpillar *Ochrogaster lunifer* Herrich-Schäffer (Lepidoptera: Thaumetopoeidae) at local and regional scales. *Australian Journal of Entomology* **38**, 330–339.

Foottit RG, Adler PH (Eds) (2009) *Insect Biodiversity: Science and Society.* Blackwell Publishing, West Sussex.

Gooderham J, Tsyrlin E (2002) *The Waterbug Book.* CSIRO Publishing, Melbourne.

Gordh G, Headrick D (2011) *A Dictionary of Entomology.* 2nd edn. CSIRO Publishing, Melbourne.

Gordon DG (1996) *The Compleat Cockroach: A Comprehensive Guide to the Most Despised (and Least Understood) Creature on Earth.* Ten Speed Press, California.

Grissell E (2010) *Bees, Wasps and Ants: The Indispensable Role of Hymenoptera in Gardens.* Timber Press, Oregon.

Guarneri F, Guarneri C, Mento G, Ioli A (2006) Pseudo-delusory syndrome caused by *Limothrips cerealium*. *International Journal of Dermatology* **45**, 197–199.

Gullan PJ, Cranston PS (2010) *The Insects: An Outline of Entomology.* 4th edn. Wiley-Blackwell, Chichester.

Guthrie DM, Tindall AR (1968) *The Biology of the Cockroach.* Edward Arnold, London.

Hackstein JHP, Stumm CK (1994) Methane production in terrestrial arthropods. *Proceedings of the National Academy of Sciences of the United States of America* **91**, 5441–5445.

Hadlington P, Marsden C (1998) *Termites and Borers: A Homeowner's Guide to Detection and Control.* UNSW Press, Sydney.

Hadlington P, Staunton I (2008) *Australian Termites.* 3rd edn. UNSW Press, Sydney.

Hadlington PW, Johnston JA (1998) *An Introduction to Australian Insects.* UNSW Press, Sydney.

Hangay G, German P (2000) *Insects of Australia.* Reed New Holland, Sydney.

Hangay G, Zborowski P (2012) *A Guide to the Beetles of Australia.* CSIRO Publishing, Melbourne.

Harvey MS, Yen AL (1989) *Worms to Wasps*. Oxford University Press, Melbourne.

Hawkeswood TJ (2005) Review of the biology and host-plants of the Australian jewel beetle *Julodimorpha bakewelli* (White, 1859) (Coleoptera: Buprestidae). *Calodema* **3**, 3–5.

Hendersen A, Hendersen D, Sinclair J (2008) *Bugs Alive: A Guide to Keeping Australian Invertebrates*. Museum Victoria, Melbourne.

Hill SR, Zaspel J, Weller S, Hansson BS, Ignell R (2010) To be or not to be…a vampire: A matter of sensillum numbers in *Calyptra thalictri*? *Arthropod Structure and Development* **39**, 322–333.

Hölldoblet B, Wilson EO (1994) *Journey to the Ants: A Story of Scientific Exploration*. Belknap Press, Cambridge, MA.

Hopkins GHE (1949) The host-associations of the lice of mammals. *Proceedings of the Zoological Society of London* **119**, 387–605.

Horne PA, Crawford DJ (1996) *Backyard Insects*. Melbourne University Press, Melbourne.

Howard DF, Blum MS, Fales HM (1983) Defense in thrips: forbidding fruitiness of a lactone. *Science* **220** (4594), 335–336.

Huang HT, Yang P (1987) The ancient cultured citrus ant. *BioScience* **37** (9), 665–671.

Imes R (1992) *The Practical Entomologist: Where and How to Look for Insects, How to Record What You See and How to Collect Them*. Aurum Press, London.

Integrated Taxonomic Information System on-line database http://www.itis.gov.

Kamimura Y (2006) Right-handed penises of the earwig *Labidura riparia* (Insects, Dermaptera, Labiduridae): evolutionary relationships between structural and behavioural asymmetries. *Journal of Morphology* **267** (11), 1381–1389.

Keller L, Gordon E (2009) *The Lives of Ants*. Oxford University Press, London.

Kirk WDJ (1996) *Thrips*. Richmond Publishing, Slough.

Kittler R, Kayser M, Stoneking M (2003) Molecular evolution of *Pediculus humanus* and the origin of clothing. *Current Biology* **13** (16), 1414–1417.

Kittler R, Kayser M, Stoneking M (2004) Erratum: molecular evolution of *Pediculus humanus* and the origin of clothing. *Current Biology* **14** (24), 2309.

Lambdin P (2005) Scale insects and mealybugs (Hemiptera: Coccoidea). In *Encyclopedia of Entomology*. (Ed JL Capinera) Part 19. Springer, Netherlands.

Lee KE, Wood TG (1971) *Termites and Soils*. Academic Press, London.

Lindsey T (1998) *Spiders of Australia*. New Holland, Sydney.

Llewellyn R (Ed.) (2002) *The Good Bug Book: Beneficial Organisms Commercially Available in Australia and New Zealand for Biological Pest Control*. Integrated Pest Management, Queensland.

Lloyd AG (1980) Extraction and chemistry of cochineal. *Food Chemistry* **5** (1), 91–107.

Lyons RE, Wong DCC, Kim M, Lekieffre N, Huson MG, Vuocolo T, Merritt DJ, Nairn KM, Dudek DM, Colgrave ML, Elvin CM (2011) Molecular and functional characterisation of resilin across three insect orders. *Insect Biochemistry and Molecular Biology* **41**, 881–890.

Marshall SA (2012) *Flies: The Natural History and Diversity of Diptera*. Firefly Books, New York.

Moran NA, Jarvik T (2010) Lateral transfer of genes from fungi underlies carotenoid production in aphids. *Science* **328** (5978), 624–627.

Moulds MS (1990) *Australian Cicadas.* UNSW Press, Sydney.

Naumann I (1993) *CSIRO Handbook of Australian Insect Common Names: Common and Scientific Names for Insects and Allied Organisms of Economic and Environmental Importance.* CSIRO Publishing, Melbourne.

New TR (1991) *Insects as Predators.* UNSW Press/Australian Institute of Biology, Sydney.

New TR (1992) *Introductory Entomology for Australian Students.* UNSW Press, Sydney.

New TR (1996) *Name that Insect: A Guide to the Insects of Southeastern Australia.* Oxford University Press, Melbourne.

Orr A, Kitching R (2010) *The Butterflies of Australia.* Jacana Books/Allen & Unwin, Sydney.

Pechenik JA (2010) *Biology of the Invertebrates.* 6th edn. McGraw-Hill, New York.

Raven R, Seeman O (2007) Arachnids. In *Wildlife of Greater Brisbane.* 2nd edn, pp. 31–67. Queensland Museum, Brisbane.

Rees D (2007) *Insects of Stored Grain: A Pocket Reference.* CSIRO Publishing, Melbourne.

Rentz D (1996) *Grasshopper Country: The Abundant Orthopteroid Insects of Australia.* UNSW Press, Sydney.

Rentz D (2010) *A Guide to the Katydids of Australia.* CSIRO Publishing, Melbourne.

Rentz D (2014) *A Guide to the Cockroaches of Australia.* CSIRO Publishing, Melbourne.

Rentz DCF, Lewis RC, Su YN, Upton MS (2003) *A Guide to Australian Grasshoppers and Locusts.* Natural History Publications, Borneau, Kota Kinabalu.

Robinson WH (2005) *Urban Insects and Arachnids: A Handbook of Urban Entomology.* Cambridge University Press, Cambridge.

Roth LM, Willis ER (1952) Tarsal structure and climbing ability of cockroaches. *Journal of Experimental Zoology* **119** (3), 483–517.

Rothschild M, Gardiner B, Mummery R (1978) The role of carotenoids in the 'golden glance' of danaid pupae (Insecta: Lepidoptera). *Journal of Zoology* **186** (3), 351–358.

Schafer R, Sanchez TV (1973) Antennal sensory system of the cockroach, *Periplaneta americana*: postembryonic development and morphology of the sense organs. *Journal of Comparative Neurology* **149** (3), 335–354.

Schmidt JO, Blum MS, Overal WL (1984) Hemolytic activities of stinging insect venoms. *Archives of Insect Biochemistry and Physiology* **1** (2), 155–160.

Schutze MK, Yeates DK, Graham GC, Dodson G (2007) Phylogenetic relationships of antlered flies, *Phytalmia* Gerstaecker (Diptera: Tephritidae): the evolution of antler shape and mating behaviour. *Australian Journal of Entomology* **46**, 281–293.

Shattuck SO (1999) *Australian Ants: Their Biology and Identification.* CSIRO Publishing, Melbourne.

Silsby J (2001) *Dragonflies of the World.* CSIRO Publishing, Melbourne.

Simon-Brunet B (1994) *The Silken Web.* Reed Books, Sydney.

Smith D, Beattie GAC, Broadley R (Eds) (1997) *Citrus Pests and Their Natural*

Enemies: Integrated Pest Management in Australia. Queensland Department of Primary Industries, Brisbane.

Surhone LM, Tennoe MT, Henssonow SF (2010) *Australian Native Bees: Bee, Wasp, Paper Wasp*. Betascript Publishing, Mauritius.

Swaine G, Ironside DA (1983) *Insect Pests of Field Crops in Colour*. Queensland Department of Primary Industries, Brisbane.

Swaine G, Ironside DA, Yarrow WHT (1985) *Insect Pests of Fruit and Vegetables in Colour*. Queensland Department of Primary Industries, Brisbane.

Terry I, Walter GH, Moore C, Roemer R, Hull C (2007) Odor-mediated push–pull pollination in cycads. *Science* **318** (5847), 70.

Theischinger G, Hawking J (2006) *The Complete Field Guide to Dragonflies of Australia*. CSIRO Publishing, Melbourne.

Thornton IWB (1964) Air-borne Psocoptera trapped on ships and aircraft. *Pacific Insects* **6** (2), 285–291.

Vane-Wright D (2003) *Butterflies*. Natural History Museum, London.

Waldbauer G (2003) *What Good Are Bugs? Insects in the Web of Life*. Harvard University Press, London.

Watson JAL, Theischinger G, Abbey HM (1991) *The Australian Dragonflies: A Guide to the Identification, Distributions and Habitats of Australian Odonata*. CSIRO Publishing, Melbourne.

Wheater CP, Read HJ (1996) *Animals Under Logs and Stones*. Richmond Publishing, Slough.

Yack JE (2004) The structure and function of auditory chordotonal organs in insects. *Microscopy Research and Technique* **63**, 315–337.

Young D (1972) Neuromuscular mechanism of sound production in Australian cicadas. *Journal of Comparative Physiology* **79** (4), 343–362.

Zborowski P, Edwards T (2007) *A Guide to Australian Moths*. CSIRO Publishing, Melbourne.

Zborowski P, Storey R (2010) *A Field Guide to Insects in Australia*. Reed New Holland, Sydney.

Further reading

The list below recommends some books that provide good general information on the most commonly encountered insect Orders in your home and garden. Further reading suggestions for specific groups of insects can be found on the information page for the Order to which they belong (Chapter 6).

The book ranking system
Throughout the book, suggested reading material has been given a ranking of between one and three 'beetles'. This is not a review of the merits of each book – it is a system to rank books based on their level of difficulty or complexity for the average reader. This helps you to choose the books most suitable for your level of background knowledge and your research requirements.

A ranking of one beetle is used for books that can be used and enjoyed by everyone. These books typically contain good general information which is easy to understand and has minimal use of scientific terms and technical jargon. These books usually feature lots of great colour photographs.

A ranking of two beetles indicates a more detailed book for people wanting to gain a better understanding of the different types of insects in Australia. Such books may contain scientific terms and higher levels of insect classification. Identification keys may be included in these books to help you identify insects to a more specific group than their Order. There are usually lots of colour photographs.

A ranking of three beetles is reserved for textbooks that are designed for students of science, biology or entomology and amateur entomologists. These books contain highly detailed information, including lots of technical terms and often complex scientific concepts. They generally do not contain many photographs, but often have lots of diagrams to illustrate key points.

General information
Brunet B (2000) *Australian Insects: A Natural History.* Reed New Holland, Sydney.
A beautifully photographed guide to Australian insects, including details on the lifecycles, diversity and survival strategies for the various Orders. ●●
CSIRO Division of Entomology (Eds) (1996) *The Insects of Australia: A Textbook for Students and Research Workers.* Melbourne University Press, Melbourne.
An extremely detailed book, compiled by some of Australia's leading entomologists. Well-illustrated dichotomous keys allow the reader to identify insects to Family level

(and in some cases, more specific groups such as Subfamilies and Tribes). A few pages of colour drawings illustrate some interesting Australian insects. ●●●

Gullan PJ, Cranston PS (2010) *The Insects: An Outline of Entomology*. 4th edn. Wiley-Blackwell, Chichester.
A very technical book detailing the major insect Orders, including good information on their general structure, morphology and physiology. Handy 'taxoboxes' at the back of the book summarise the features of each Order at a glance. ●●●

Hadlington PW, Johnston JA (1998) *An Introduction to Australian Insects*. UNSW Press, Sydney.
A very thorough introduction to the different Orders of insects in Australia. There are plenty of black and white drawings and photographs, as well as colour plates of some commonly encountered insect species. ●●

Hangay G, German P (2000) *Insects of Australia*. Reed New Holland, Sydney.
A simple, comprehensive guide to Australian insect Orders, featuring lots of colour photographs. ●●

Horne PA, Crawford DJ (1996) *Backyard Insects*. Melbourne University Press, Melbourne.
A great compilation of stunning colour photographs introducing some insects and arthropods commonly found in homes and gardens. ●

New TR (1992) *Introductory Entomology for Australian Students*. UNSW Press, Sydney.
A very thorough guide to Australian insect Orders, featuring lots of black and white illustrations. Of particular use are tables grouping insect Orders by the habitats in which they are most commonly found. ●●

Zborowski P, Storey R (2010) *A Field Guide to Insects in Australia*. Reed New Holland, Sydney.
A simple, comprehensive guide to Australian insect Orders, featuring lots of beautiful colour photographs. There is also a handy dichotomous key to insect Orders. ●●

Specific habitats
Inside the home

Rees D (2007) *Insects of Stored Grain: A Pocket Reference*. CSIRO Publishing, Melbourne.
A handy pocket guide to the major pests of stored cereals, grains and legumes. There is detailed information on the biology, distribution and economic importance of each species as well as full colour photographs. ●

Robinson WH (2005) *Urban Insects and Arachnids: A Handbook of Urban Entomology*. Cambridge University Press, Cambridge.
A detailed guide to cosmopolitan species of insects and arachnids that invade our homes and backyards. Contains black and white drawings of commonly encountered species. ●●

In water

Gooderham J, Tsyrlin E (2002) *The Waterbug Book*. CSIRO Publishing, Melbourne.
A beautifully photographed guide to freshwater invertebrates, including insects. Includes an easy-to-use identification key to sort aquatic invertebrates into different groups, plus more detailed keys for each Order that

can be used, with the aid of a magnifier/microscope, to sort insects more specifically into Families. ●●

In soil, leaf litter and compost

Wheater CP, Read HJ (1996) *Animals Under Logs and Stones.* Richmond Publishing, Slough.
This book contains identification keys which can be used, with the aid of a magnifier/microscope, to identify invertebrates found underneath stones, logs and leaf litter in your garden. It features detailed black and white diagrams and several pages of colour illustrations. ●●

On citrus trees

Smith D, Beattie GAC, Broadley R (Eds) (1997) *Citrus Pests and Their Natural Enemies: Integrated Pest Management in Australia.* Queensland Department of Primary Industries, Brisbane.
This book catalogues the damage caused by and the biology of over 100 citrus pests in Australia. Distribution maps and highly detailed colour photographs makes this a great reference book. ●●

In the vegetable garden

Llewellyn R (2002) *The Good Bug Book.* Integrated Pest Management, Australia.
Great information on insects (e.g. lacewings) used in the biological control of crop pests. ●

Swaine G, Ironside DA (1983) *Insect Pests of Field Crops in Colour.* Queensland Department of Primary Industries, Brisbane.
For each type of plant, this book lists the insect species that are known to attack it. It includes colour photographs, as well as information on their appearance, biology and natural enemies. ●

Swaine G, Ironside DA, Yarrow WHT (1985) *Insect Pests of Fruit and Vegetables in Colour.* Queensland Department of Primary Industries, Brisbane.
For each type of fruit or vegetable, this book lists the insect species that are known to attack it. It includes colour photographs, as well as information on their appearance, biology and natural enemies. ●

In and around flowers

Dollin A, Batley M, Robinson M, Faulkner B (2000) *Native Bees of the Sydney Region: A Field Guide.* Australian Native Bee Research Centre, North Richmond.
This easy-to-read field guide has great colour plates, descriptions and actual size scales for many of the native bee species found throughout Australia (including Sydney). Great info on how to attract native bees to your garden. ●

Non-insect arthropods

Beccaloni J (2009) *Arachnids.* CSIRO Publishing, Melbourne.
A beautifully photographed introduction to arachnids (spiders, scorpions, ticks and mites) of the world. Tables and diagrams help to simplify tricky classification and illustrate key morphological features of each group. ●●

Harvey MS, Yen AL (1989) *Worms to Wasps.* Oxford University Press, Melbourne.
A detailed guide to the different Orders of land-dwelling invertebrates in Australia, including a handy illustrated identification key. This book contains

excellent black and white illustrations and general information on Australian invertebrates, but some of the classifications have been revised since it was published. ●●

Hendersen A, Hendersen D, Sinclair J (2008) *Bugs Alive: A Guide to Keeping Australian Invertebrates*. Museum Victoria, Melbourne.
Even if keeping arthropods as pets isn't your cup of tea, this book contains amazing close-up photographs and easy-to-read information on the biology of spiders, scorpions, millipedes and centipedes, including detailed information on some of the more commonly encountered species. ●

Lindsey T (1998) *Spiders of Australia*. New Holland, Sydney.
A beautifully photographed, easy-to-read guide on the biology of Australian spiders, including information on some of the more commonly encountered species. ●

Simon-Brunet B (1994) *The Silken Web*. Reed Books, Sydney.
A beautifully photographed book on the biology of Australian spiders, including colour diagrams of some of the most commonly encountered species. ●●

Index

abdomen 5, 9, 10, 11
Aconophora compressa 142
Acrida conica 234
Acropyga sp. 177
Afranthidium repetitum 199
Agrianome spinicollis 207
alderflies 20
Alotartessus iambe 182
Amblypelta lutescens 297
Amegilla sp. 226
Amenia sp. 287
Amerila crokeri 206
ametabolous metamorphosis *see* simple metamorphosis
Amphipoda *see* land hoppers
Amphirhoe decora 219
Anisops sp. 98
Anisoptera *see* dragonflies
Anoplognathus porosus 88
antennae 5, 7, 9, 10, 11–12
Anthela varia 205
Anthrenus (Florilinus) museorum 78
anting 249
antlions 61, 153, 243, 244
  Myrmeleon sp. 153
ants 12, 14, 16, 47, 48, 58, 153, 155, 162, 163, 167, 174–83, 189, 191, 249, 263–4, 266, 271, 282–3, 297
  Acropyga sp. 177
  ant-loving insects *see* myrmecophiles
  blue 'ant' 130
  bulldog 174, 175, 178
  bullet 178–9
  Camponotus perthiana 178
  citrus 179
  coastal brown 182
  dome-backed spiny 155
  fire 178
  giant bull 266
  golden-tailed spiny 182
  green 131
  greenhead 131, 178
  green tree 178, 179
  jumping 178
  larvae 68, 155, 177
  meat 132, 176
  pasture funnel 153–4, 178
  queen 176, 177, 178, 179
  soldier 176, 177
  spider 181
  sugar 271
  suicide bomber 178
  weaver 179
  white *see* termites
  winged 176, 177
  worker 163, 176, 177, 179, 182
Aonidiella aurantii 151
Aphaenogaster pythia 154
Aphididae *see* aphids
aphids 13, 18, 65, 106, 150, 176, 194, 245, 271, 274–5, 277, 278, 279, 280, 281, 282, 283, 284
  black citrus 245
  green peach 271
  onion 194
  pea 284
  rose 106
Apis mellifera 164, 165, 166, 167, 286
Apocrita *see* ants; bees; wasps
Apoidea *see* bees
appendages *see* legs; wings
aquatic insects 20, 26, 67, 93–101, *see also* water bugs
arachnids *see* mites; scorpions; spiders; ticks
Archaeognatha *see* bristletails
Archimantis latistyla 260
arthropods 5, 8, 27, 29, 297–303
assassin bugs 109, 277, 294, 295, 296, 297
  bee killer 7, 109
  feather-legged 297
Atractomorpha similis 124
Auchenorrhyncha 275, 277, 285, 287, *see also* cicadas; hoppers; spittle bugs
Austroaeschna 99
  unicornis 218

Bactrocera cucumis 169, 232
Balta sp. 187
barklice *see* booklice
bees 12, 13, 14, 16, 32, 162, 163–8, 198–9, 204, 226, 286
  bluebanded 165, 226
  carder 199
  carpenter 164, 165
  cuckoo *see* cuckoo bees
  honey 164, 165, 166, 167, 286
  larvae 68
  leafcutting 147–8, 165
  mason 165
  Megachile monstrosa 173
  queen 165–6
  resin 168
  solitary 165, 166, 168
  stingless 164, 165, 166, 167, 168
  worker 165, 167
beetles 13, 14, 16, 30, 31, 34, 35, 51, 63, 147, 148, 149, 186–94, 207, 219, 293
  ant-loving 191

bombardier   190
cane   *see* cane beetles
carpet   78
carrion   189
chafer   87
Christmas   6, 69, 71, 88, 188, 189, 190
cigarette   74
click   69
cockchafer   103
cowboy   187
curl grubs   69, 103
darkling   191
diving   *see* diving beetles
dung   69, 189, 191
fiddler   293
flower   134
ground   189
grubs   51, 69, 190
jewel   *see* jewel beetles
ladybird   *see* ladybird beetles
ladybugs   *see* ladybird beetles
larvae   69, 147, 148, 190
leaf   *see* leaf beetles
longicorn   *see* longicorn beetles
mealworm   191
museum   78
pie-dish   190
rhinoceros   69, 86, 103, 189, 190, 191, 192
rove   34, 189
scarab   *see* scarab beetles
stag   *see* stag beetles
weevil   *see* weevils
whirligig   94
Belenois java   140
Belostomatidae   *see* water bugs, giant
Bemisia tabaci   275
beneficial insects, examples of   166–7, 191, 224, 296
biological control, examples of   92, 142, 172, 178, 179, 191, 192, 202, 244–5, 254

Biprorulus bibax   117
bites   84, 95, 97, 98, 100, 109, 111, 135–7, 223, 299, *see also* itchy hairs; stings
on leaves   147
Blattaria   *see* cockroaches
Blattella germanica   79, 209
Blattellidae   *see* cockroaches, bush
Blattodea   *see* cockroaches; termites
bloodworms   101
blowflies   226, 228, 231, 232
  hairy maggot   231
  larvae   68
  snail parasitic   287
Bombyx mori   37
booklice   11, 14, 16, 64, 194–7, 247, 278–9
  Ectopsocus sp.   194
  Liposcelis sp.   247
  Myopsocus sp.   197
  Propsocus pulchripennis   279
Brachycera   226, 228, *see also* flies
bristletails   19, 22
bugs   7
  assassin   *see* assassin bugs
  banana-spotting   297
  bed   65, 84, 294, 296, 297
  bronze orange   22, 116, 295
  brown bean   123, 180
  cotton harlequin   108
  crusader   14, 208, 277
  green vegetable   122, 296
  jewel   108, 277, 294, 295
  moss   277
  plant   63, 188, 276, 277, 292, 293, 294, 296
  pod-sucking   123, 295
  shield   *see* shield bugs
  spined citrus   117
  stink   13, 18, 63, 108, 116, 117, 126, 188, 276, 277, 292, 293, 295

water   *see* water bugs
butterflies   12, 13, 15, 16, 37, 71, 197–206, *see also* caterpillars
  black jezebel   198
  blue tiger   141
  cabbage white   120, 201, 202
  Cairn's birdwing   12, 15, 206
  caper white   140
  chrysalis   23, 200, 201
  common crow   157–8
  glasswing   37
  larvae   *see* caterpillars
  monarch   37, 205
  skippers   201
  swallowtail   *see* swallowtail butterflies
  tailed emperor   23
  wanderer   37, 205

caddisflies   20
Caedicia sp.   91
Caelifera   233–4, 236, *see also* grasshoppers; locusts
Calofulcinia sp.   255
Calyptra sp.   202
camouflage   91, 97, 99, 118, 120, 146, 147, 190, 201, 209, 216, 237, 254, 255, 261, 262, 263, 295
Campion sp.   253
Camponotus   271
  perthiana   178
cane beetles   69, 87, 103, 190
  greyback   87
case moths   155–7
  Saunder's   156
castes   60, 175, 176, 266–7
caterpillars   13, 23, 51, 66, 69–70, 147, 148, 155, 171, 172, 185–6, 200, 201, 202
  hairy   69–70, 202, 205
  inch worm   69–70
  itchy grub   205
  lawn armyworm   128, 129

processionary   138–9
swallowtail   118, 201
centipedes   29, 64, 66, 67, 297, 300–1
   Scolopendra laeta   301
Cephonodes kingii   104, 204
Ceratopogonidae   *see* midges, biting
cerci   9, 10, 11, 36, 53
Cercophonius sp.   299
Ceroplastes rubens   152
Charaxes sempronius   23
Cheleutoptera   *see* stick insects
cherry slugs   5, 186
Chironomus sp.   101
Chlorobalius leucoviridis   240
Chondropyga dorsalis   187
Chortoicetes terminifera   143
chrysalis   23
Chrysomelidae   *see* leaf beetles
Chrysomya sp.   232
cicadas   11, 13, 18, 43, 107, 227, 242, 275–6, 277, 285–92
   bladder   288
   bottle   285
   brown bunyip   242
   Dodd's bunyip   107
   razor grinder   227
   shells   107, 289
Cicadidae   *see* cicadas
Cimex lectularius   65, 84, 294, 296, 297
cochineal insects   284
cockroaches   11, 13, 14, 16, 19, 49, 62, 159, 187, 207–12, 256–7, 268
   'albino'   209
   American   79–80, 210, 257
   Australian   80
   balta   187
   burrowing   105
   bush   110
   ellipsidion   110, 207
   German   79, 209
   giant burrowing   210, 211
   household   159, 209, 210
   oothecae   79, 159, 209
   Sloane's northern wingless   211
   wood   105
cocoon   23, 155
Coleoptera   *see* beetles
Coleorrhyncha   277
Collembola   *see* springtails
Coloburiscoides sp.   67
colony   *see* nests
complete metamorphosis   23
compound eyes   9, 10
Corixidae   *see* water bugs, water boatman
Coscinocera hercules   198
Cosmozosteria sloanei   211
crickets   11, 13, 14, 17, 38–9, 54–5, 233–40, 252, 260, *see also* katydids
   ant   236
   bark   239
   bush   236
   field   *see* field crickets
   king   236, 237
   mole   *see* mole crickets
   Nullarbor cave   238
   raspy   236, 237
Crustaceans   *see* land hoppers; slaters
Cryptolaemus montrouzieri   192
Ctenocephalides felis   222, 223, 224, 247
Ctenolepisma   21
   longicaudata   256
cuckoo bees   167
   neon   167
Culicidae   *see* mosquitoes
Culicoides sp.   136
Curculionidae   *see* weevils
Cycadothrips chadwicki   270, 273–4

damselflies   14, 16, 45, 99, 212–17, 241–2
   bluetail   99
   bronze needle   213
   common bluetail   242
   larvae   99, 215, 216
   naiads   99, 216
   nymphs   99, 216
Danaus plexippus   37, 205
Delias nigrina   198
Dermaptera   *see* earwigs
Dermolepida albohirtum   87
Derocephalus angusticollis   232
Diamma bicolor   130
Dictyoptera   *see* cockroaches; praying mantids
Diphucephala sp.   134
Diplonychus sp.   296
Diplura   *see* diplurans
diplurans   19, 36, 53
Diptera   *see* flies
disease transmission by insects
   animal diseases   100, 135, 223–4, 230, 231, 249, 300
   plant diseases   106, 272, 283, 290, 296
diving beetles   93, 189
   green   93
Dolichopodidae   *see* flies, longlegged
dragonflies   11, 13, 14, 16, 22–3, 41, 45, 99, 212–18, 241–2
   darner   99
   fiery skimmer   218
   jewel flutterer   217
   larvae   22–3, 99, 215, 216
   naiads   22–3, 99, 216
   nymphs   22–3, 99, 216
   painted grasshawk   212
   unicorn darner   218
   variable tigertail   37
Drosophila sp.   77
Dytiscidae   *see* diving beetles

earwigs 13, 14, 17, 31, 36, 53, 187–8, 219–21
  common brown 219, 221
  European 187, 220
ecdysis *see* moulting
Ectobiidae *see* cockroaches, bush
Ectopsocus sp. 194
Ellipsidion 110
  reticulatum 207
Embiidina *see* web-spinners
Embioptera *see* web-spinners
Ensifera 234, 236, *see also* crickets; katydids
Ephemeroptera *see* mayflies
Ephippityha
  kuranda 252
  trigintiduoguttata 2
Epiprocta *see* dragonflies
Epiproctophora *see* dragonflies
Eulophidae *see* wasps, eulophid
Euploea core 157–8
Eupoecila australasiae 293
Eurycnema goliath 264
Eusynthemis aurolineata 37
Eutryonia monstrifer 290
exoskeleton 5, 20–2
explosive defacation 201
Extatosoma tiaratum 21, 41, 180, 251, 263
eyes *see* compound eyes
eye-spots *see* false eyes

false eyes 82, 201
field crickets 90
  mottled 90
Filipalpia *see* stoneflies
fleas 17, 64, 221–5, 247
  cat 222, 223, 224, 247
  dog 223
  echidna 224
  European rabbit 224
  human 223
  rat 224
  southern kiore 224
flies 11, 13, 14, 17, 32, 68, 149, 150, 164, 169, 183–4, 225–33, 279–80, 286–7, *see also* midges; mosquitoes
  antlered 230
  banana stalk 232
  bat 230–1
  bird 230–1, 249
  blow *see* blowflies
  bush 230
  crane 113, 225, 228
  cucumber 169, 232
  fairy 'fly' 172
  flesh 81, 228
  fly strike 231
  fruit 77, 228, 229, 230, *see also* flies, vinegar
  fungus gnat 231
  green 112
  horse 137, 226, 228
  house 13, 226, 228, 230
  hover *see* hover flies
  larvae 68, 229–30, 232, *see also* flies, maggots
  longlegged 112, 229
  louse 230–1, 249
  maggots 51, 68, 228–9, 230, 231
  March 137, 228
  moth 83, 228
  robber 228, 229, 230, 231, 232
  sand 101, 136, 228
  soldier 68, 102
  tachinid 122, 172, 226, 229, 230
  vinegar 77
Forficula auricularia 187, 220
Formicidae *see* ants
Frenatae *see* butterflies; moths
froghoppers 277, 288, 289, 291

galls 150, 170, 272, 281
Gasteruptiidae *see* wasps, gasteruptiid
Gerridae *see* water bugs, water striders
Glaucopsaltria viridis 285
glow worms 231
grasshoppers 11, 13, 14, 17, 38–9, 54–5, 147, 233–40, 252, 260, *see also* locusts
  chameleon 238
  fireman 238
  hedge 111, 236
  longheaded 234
  long-horned *see* Ensifera
  spotted bandwing 239
  vegetable 124
growth 20
Grylloptera *see* crickets; katydids
Gryllotalpa sp. 89
Gyrinidae *see* beetles, whirligig

hawk moths
  bee 37, 104, 204
  larvae 69–70, 104
  pupae 104
head 5, 9, 10
Hednota grammellus 143
Heirodula majuscula 254, 255
Helicoverpa
  armigera 121
  punctigera 121
Heliothis see Helicoverpa
hemimetabolous metamorphosis *see* incomplete metamorphosis
Hemiptera 7, 13, 18, 154, 161, 274, 277, 293, 294
  Auchenorrhyncha 275, 277, 285, 287, *see also* cicadas; hoppers; spittle bugs

Heteroptera   14, 63, 276, 277, 292, 293, 294, *see also* assassin bugs; bugs; shield bugs; water bugs
  Sternorrhyncha   274–5, 277, 278, 280, *see also* aphids; lerp insects; mealybugs; psyllids; scale insects; whiteflies
Henicopsaltria eydouxii   227
Hermetia illucens   102
Hesthesis sp.   193
Heteroptera   14, 63, 276, 277, 292, 293, 294, *see also* assassin bugs; bugs; shield bugs; water bugs
Hiarchas angularis   276
hives *see* nests
Holognatha *see* stoneflies
holometabolous metamorphosis *see* complete metamorphosis
Homoptera *see* Auchenorrhyncha; Sternorrhyncha
honeydew   106, 132, 142, 151, 152, 176–7, 182, 281, 282, 285, 288, 289, 291
hoppers   285–92, *see also* froghoppers; leafhoppers; planthoppers, treehoppers
  passionvine   291
hover flies   33, 133, 228, 229, 230
  black-headed   133, 184
  wasp-mimicking   164
Hyalinaspis sp.   151
Hymenoptera *see* ants; bees; sawflies; wasps

Icerya purchasi   284
Idolthrips spectrum   273
incomplete metamorphosis   22
insects   5, 8

lifecycles *see* growth; larvae; metamorphosis; moulting
  Orders   12–13, 19–20
  species numbers   6, 161
  taxonomy   11–12, 15
invertebrates   5, 8
Iridomyrmex sp.   132
Ischnura   99
  heterosticta   242
Isopoda *see* slaters
Isoptera *see* termites
itchy hairs   69, 138, 201, 202, 205
Ithone sp.   199
Ixodes sp.   300

jewel beetles   188, 189, 191, 192
  Julodimorpha bakewelli   191
Jugatae *see* butterflies; moths
Julodimorpha bakewelli   191
jumping plant lice *see* psyllids

katydids   91, 234, 236, 237, 238, 252, 260
  ant-mimicking   180
  Caedicia sp.   91
  Kuranda spotted   252
  mottled   2
  nicsara   261
  predatory *see* predatory katydids
  slender bush   234
  stick   39
Kiefferulus sp.   280

Labidura truncata   219, 221
Laccotrephes tristis   97
lacewings   12, 14, 17, 43, 159, 213–14, 241–6
  antlion *see* antlions
  blue eyes   159, 214
  brown   243, 244, 245
  eggs   159

  green   92, 159, 243, 244, 245
  larvae   59, 61, 244, 245
  mantis flies *see* mantis flies
  mantispids   40
  moth   199, 243
  owl flies *see* owl flies
  spongeflies   244
  Stenosmylus sp.   241
  stink flies   244
ladybird beetles   69, 190, 191
  mealybug   192
Lamprima latreillii   192
land hoppers   181, 302–3
larvae   23, 29, 51, 64, 66–70
Lasioderma serricorne   74
leaf beetles   69, 114, 148, 189
  eucalypt   114
  figleaf   148
leafhoppers   43, 227, 242, 275–6, 277, 285–92
  brown   182
  Rosopaella cf. *crofta*   275
  Stenocotis depressa   290
leaf insects   261, *see also* stick insects
  spiny   21, 41, 180, 251, 263
legs   5, 9, 10, 27, 29
Lepidogryllus sp.   90
Lepidoptera *see* butterflies; moths
Leptomyrmex sp.   181
Leptopius sp.   115
lerp insects   151, 281
  Hyalinaspis sp.   151
Lethocerus (*Lethocerus*) *insulanus*   95
lice   17, 64, 135, 195, 222, 246–50
  bird   195, 222
  biting   248
  body   248
  dogbiting   249

head   247, 248, 249
human   248, 249
rabbit   249
sucking   248, 249
lifecycles *see* metamorphosis
Liposcelis sp.   247
Lissopimpla excelsa   129
locusts   233, 236, 237, 238,
   see also grasshoppers
   Australian plague   143
longicorn beetles   34, 189,
   190
   Amphirhoe decora   119
   poinciana   207
   wasp-mimicking   193
losing limbs   27, 237, 262
Lophyrotoma   186
   interrupta   183

Macrolepidoptera *see*
   butterflies; moths
Macropanesthia
   rhinoceros   210, 211
Macrosiphum (Macrosiphum)
   rosae   106
maggots *see* flies, larvae
maggot therapy   231
Mallada signatus   92, 159,
   243, 244, 245
mantids *see* praying mantids
mantis flies   40, 242, 243,
   244, 245, 252–3
   Campion sp.   253
mantispids *see* mantis flies
Mantodea *see* praying
   mantids
Mastotermes
   darwiniensis   175, 265,
   269
mayflies   20
   spiny   67
mealybugs   51, 150, 152–3,
   176–7, 274–5, 277, 278,
   280, 281, 282, 283
   Monophlebulus sp.   283
   solenopsis   153

Mecoptera *see* scorpion
   flies
Megachile   168
   monstrosa   173
Megachile (Eutricharaea)
   sp.   148
Megaloptera *see* alderflies
Melangyna sp.   133, 184
Mesembrius sp.   164
metamorphosis   22–3
Metura elongata   156
Microlepidoptera *see*
   butterflies; moths
Mictis profana   14, 208
midges   215, 225, 228
   biting   136
   bloodworm   101
   non-biting   101, 280
migration   121, 138, 140,
   141, 143
millipedes   5, 29, 64, 66–7,
   297, 301–2
mimicry   33, 118, 179–80,
   185, 193, 199, 201, 244,
   290, *see also* camouflage
minibeast   5
mites   150, 297, 300
   red velvet   300
mole crickets   38, 89, 130,
   236, 237
   Gryllotalpa sp.   89
   pygmy   236
Monophlebulus sp.   283
mosquitoes   13, 33, 100,
   135, 215, 216, 225, 228,
   229, 230
   giant   225
   grey striped   135
   wrigglers   68, 99, 100,
   216, 229, 230, 296
moths   12, 13, 16, 37, 51,
   149, 197–206, 243, *see also*
   caterpillars
   anthelid   205
   bag-shelter   138
   cactoblastis   202

case *see* case moths
citrus leafminer   149
clothes   85
corn earworm   121
Croker's frother   206
cup   69–70
dunny   82
emperor   199
emperor gum   200
fruit-piercing   201
goat   203
Granny's cloak   82
hawk *see* hawk moths
heliothis   121
Hercules   198
Indian meal   76
larvae *see* caterpillars
native budworm   121
pantry   74, 76
pasture webworm   243
scribbly gum   149–50
silk   37
swift   199
tiger   202
vampire   202
moulting   20–2
mouthparts   7, 9, 10, 11, 13
Musgraveia sulciventris   22,
   116, 295
Myopsocus sp.   197
Myrmarachne sp.   179
Myrmecia   174
   gulosa   266
myrmecomorphy   179–80
myrmecophiles   191
Myrmeleon sp.   153
myxomatosis   224, 249
Myzus persicae   271

naiads   23, 99, 215, 216
Nasutitermes exitiosus   269
Nematocera   225, 228, *see
   also* flies, crane; midges;
   mosquitoes
Nephrotoma
   australasiae   113

Nepidae *see* water bugs, needle bugs and water scorpions
Neomantis australis 242
Neotoxoptera formosana 194
Nephila pilipes 298
nests
  ant 131, 132, 153–4, 155, 175, 176, 177, 178, 179, 263–4
  bee 165, 166, 167, 168
  termite 266, 267
  wasp 154–5, 170, 171, 172, 173, 174
Neuroptera *see* lacewings
Neurothemis stigmatizans 212
Nezara viridula 122, 296
Nicsara sp. 261
nits *see* lice
non-insect arthropods *see* centipedes; land hoppers; millipedes; mites; scorpions; slaters; spiders; ticks
Notonectidae *see* water bugs, backswimmers
Nymphes myrmeleonides 159, 214
nymphs 22–3

ocelli 9, 10
Ochlerotatus vittiger 135
Ochrogaster lunifer 138–9
Odonata *see* damselflies; dragonflies
odour, offensive 92, 178, 190, 201, 206, 209, 237, 244, 264, 295, *see also* bugs, stink
Oechalia schellenbergii 292
Ogmograptis sp. 150
Ommatius sp. 232
Onychohydrus scutellaris 93
oothecae
  cockroach 79, 159, 209

grasshopper 236–7
praying mantis 157–9, 254
Ornithoptera euphorion 12, 15, 206
Orthetrum villosovittatum 218
Orthodera ministralis 235
Orthoptera *see* crickets; grasshoppers; katydids; locusts
osmeteria 201, 206
owl flies 243
  Suhpalacsa cf. *lyriformis* 246

Panesthia cribrata 105
paper wasps 68, 172, 174
  Polistes sp. 174
  yellow 162
Papilio (Princeps) aegeus 119, 202
  fuscus 118
Paropsis maculata 114
parthenogenesis 248, 262, 272, 282
pear slugs 186
Pediculus humanus 248, 249
  capitus 247, 248, 249
  humanus 248
Perga 127
  dorsalis 162
Pergidae *see* sawflies, spitfires
Periplaneta americana 79–80, 210, 257
Periplaneta australasiae 80
Phaneroptera gracilis 234
Phasmatodea *see* stick insects
Phasmatoptera *see* stick insects
Phasmida *see* stick insects
Pheidole megacephala 182
Phenacoccus solenopsis 153
pheromones 167, 175, 178, 200, 229, 244, 262

Philomastix macleaii 186
Philopteridae *see* lice, bird
Phthiraptera *see* lice
Phylacteophaga sp. 148–9
Phyllocnistis citrella 149
Pieris rapae 120, 201, 202
pillbugs *see* slaters
plagues 134, 138
  bubonic 223
  locust 143
Planipennia *see* lacewings
planthoppers 227, 242, 275–6, 277, 285–92
  eucalypt 125
Platybrachys sp. 125
Plecoptera *see* stoneflies
Plodia interpunctella 76
Podacanthus viridiroseus 235
Poecilometis 126
  histricus 188
poisonous insects 190, 201, 205, 209, 237
Polistes 174
  dominulus 162
pollination 104, 120, 129, 132, 133, 134, 148, 166–7, 202, 230, 273–4
polyembryony 171
Polyrachis sp. 182, 155
Pompilidae *see* wasps, spider
Poneridia australis 148
praying mantids 11, 13, 14, 18, 40, 56, 157–9, 235, 242, 250–6, 259–60
  burying 250
  Calofulcinia sp. 255
  garden 235
  giant green 254, 255
  netwinged 242
  oothecae 157–9, 254
  stick 260
predatory katydids 38
  spotted 240
Pristhesancus plagipennis 7, 109

prolegs 29, 66–7
Protura *see* proturans
proturans 19
psocids *see* booklice
Psocodea *see* booklice; lice
Psocoptera *see* booklice
Propsocus pulchripennis 279
Psychoda sp. 83
Psychodidae *see* flies, moth
psyllids 150, 281
Ptilocnemus lemur 297
Pulvinaria urbicola 192
pupae 23

Qualetta maculata 239

Ranatra dispar 97
Reduviidae *see* assassin bugs
resilin 224–5, 237
Rhyothemis resplendens 217
Rhytidoponera metallica 131
Riptortus serripes 123, 180
Rosopaella cf. *crofta* 275
Rutilia (Donovanius)
  sp. 226

Sarcophagidae *see* flies, flesh
sawflies 14, 16, 149, 162,
  163, 183–6
  cattle-poisoning 183
  cherry slugs 5, 186
  larvae 66, 68–9, 127,
    147, 148, 149, 184, 185,
    186
  leafblister 148–9, 184
  Lophyrotoma sp. 186
  pear slugs 186
  Perga sp. 127
  raspberry 186
  spitfires 68, 127, 185
  steelblue 162
  wood wasp 184–5
scale insects 51, 274–5, 277,
  278, 280, 281, 282, 283,
  284
  armoured 151–2

California red 151
cochineal 284
cottony cushion 284
cottony urbicola 192
hard 151–2
pink wax 152
soft 152
Scarabaeidae *see* scarab
  beetles
scarab beetles 69, 103, 191
  green 134
Sceliphron sp. 155
Schmidt pain index 166,
  172, 178
Scolopendra laeta 301
Scolypopa australis 291
scorpion flies 19
scorpions 5, 297, 299
  wood 299
setae 9, 10
Setipalpia *see* stoneflies
shedding *see* moulting
shield bugs 126, 177, 294,
  295
  gum tree 126
  Hiarchas angularis 276
  spined predatory 292
  zebra gum tree 188
short-horned grasshoppers
  see grasshoppers; locusts
silverfish 18, 21, 22, 52,
  256–9
  Ctenolepisma sp. 21
  gray 256
simple metamorphosis 21,
  22
Siphonaptera *see* fleas
Sipyloidea caeca 259
Sitophilus oryzae 75
skin *see* exoskeleton
slaters 302
social insects *see* ants; bees;
  termites; wasps
soft bugs *see* Sternorrhyncha
Speiredonia spectans 82
Sphingidae *see* hawk moths

Sphodropoda tristis 250
spiders 5, 29, 154–5, 172,
  173, 174, 245, 297, 298–9
  ant-mimicking 179
  golden orb weaver 298
  trapdoor 173
spiracles 10, 11
Spirobolida *see* millipedes
spitfires 68, 127, 185
spittle bugs 154, 287, 289
Spodoptera sp. 128
springtails 19
stag beetles 108, 189
  golden green 192
Stenocotis depressa 290
Stenosmylus sp. 241
Sternorrhyncha 274–5, 277,
  278, 280, *see also* aphids;
  lerp insects; mealybugs;
  psyllids; scale insects;
  whiteflies
stick insects 11, 14, 18, 41,
  57, 147, 235, 251–2, 259–
  65, *see also* leaf insects
  confused winged 259
  goliath 264
  Kirby's 264
  Lord Howe island 263
  peppermint 264
  red-winged 235
  spiny leaf 21, 41, 180,
    251, 263
  titan 263
stings 130, 131, 162, 163,
  166, 167, 169, 170, 171,
  172, 176, 177–8, 299, *see
  also* itchy hairs
stoneflies 20
Stratiomyidae *see* flies, soldier
Strepsiptera *see* stylops
stridulation, examples
  of 86, 185, 209, 237,
  262, 295
stylops 19
Suhpalacsa cf. *lyriformis* 246
swallowtail butterflies 118

clearwing 37
fuscous 118
larvae 118, 201
orchard 119, 202
symbiosis, examples of
   ants and sap-sucking
      bugs 152, 176–7, 182,
      282–3, 288, 289
   ants and stick insect
      eggs 263
   beetles and ants 191
   cockroaches and
      protozoa 209
   termites and
      protozoa 267
   thrips and cycads 273–4
Symphyta *see* sawflies
Synlestes weyersii 213
Syrphidae *see* hover flies

Tabanidae *see* flies, horse
Tamasa
   doddi 107
   tristigma 242
Tectocoris diopthalmus 108
Tenagogerris euphrosyne 96
termites 13, 14, 18, 19, 48,
   51, 59–60, 175, 177, 189,
   258, 265–70
   alates 267
   giant northern 175, 265,
      269
   king 267
   Nasutitermes
      exitiosus 269
   queen 51, 267, 269
   reproductives 267
   soldier 59, 60, 267–8,
      269
   worker 59, 60, 267, 269
Tetragonula carbonaria 168
Tettigoniidae *see* katydids
Theseus sp. 126
thorax 5, 9,10
thrips 18, 65, 270–4
   cycad 270, 273–4

giant 273
grain 273
Thyreus nitidulus 167
Thysanoptera *see* thrips
Thysanura *see* silverfish
ticks 5, 297, 300
   Ixodes sp. 300
Tineola sp. 85
Tipulidae *see* flies, crane
Tirumala hamata 141
Torbia sp. 180
Toxoptera citricida 245
Toxorynchites speciosus 225
treehoppers 288, 289, 291
   horned 290
   lantana 142
Trialeurodes
   vaporariorum 278
Trichoptera *see* caddisflies
Trombidiidae *see* mites
true bugs *see* Hemiptera
typical bugs *see* Heteroptera

Valanga irregularis 111, 236
vectors *see* disease
   transmission by insects
venom 163, 166, 171, 172,
   174, 299
   claws 301
viruses *see* disease
   transmission by insects

warning colouration,
   examples of 171, 185,
   190, 193, 237, 244, 295
wasps 11, 14, 16, 32, 68,
   149, 150, 162, 163, 165,
   167, 169–74, 198–9, 226,
   286
   blue-'ant' 130
   chalcidoid 170
   cuckoo 170
   encyrtid 171
   eulophid 173
   European 172
   fairy 'fly' 172

flower 172
gall-forming 150, 170
gasteruptiid 173
hornet 167, 172
larvae 68, 170, 171
mason 154
mud dauber 154, 155
orchid dupe 129
paper *see* paper wasps
parasitic 159, 170, 171,
   172, 284
potter 154, 155
predatory 170, 171
queen 171
social 171
solitary 170
spider 170, 172, 173, 174
trigonalid 172
wood 'wasp' 184
worker 171
yellow-banded
   ichneumon 169
water bugs 63, 188, 276,
   277, 292, 293, 294, 295,
   296
   backswimmers 98, 277,
      294
   giant 95, 277, 294, 295,
      296
   needle bugs 97, 295
   pond skaters 96
   toe-biters 95
   water boatman 98
   water scorpions 97, 277,
      295, 297
   water striders 96, 294,
      295
web-spinners 19
weevils 115, 189, 190, 191,
   192
   broad-nosed 115
   rice 75
   salvinia 191
whiteflies 277, 278, 279,
   280, 281, 282, 283
   glasshouse 278

silverleaf   275
white grubs *see* beetles, curl grubs
wings   9, 10, 11, 14, 22, 23, 31, 32
  buds   22, 23, 31

witchetty grubs *see* beetles, curl grubs
woodlice *see* slaters

Xanthopimpla rhopaloceros   169

Xeroderus kirbii   264
Xylotrupes ulysses   86, 103, 192

Zaclotathra oligoneura   239
Zygentoma *see* silverfish
Zygoptera *see* damselflies